The Estate House Re-designed

Anthony Sully

The Estate House Re-designed

Springer

Anthony Sully
SullyDesigns
Ely, Cambridgeshire, UK
e-mail: sultony@gotadsl.co.uk

ISBN 978-3-030-08006-8 ISBN 978-3-319-90397-2 (eBook)
https://doi.org/10.1007/978-3-319-90397-2

Softcover re-print of the Hardcover 1st edition 2019

Printed on acid-free paper

This Springer imprint is published by the registered company Springer Nature Switzerland AG
The registered company address is: Gewerbestrasse 11, 6330 Cham, Switzerland

Foreword

Buildings and the demands placed on them are constantly changing, yet the industry's approaches and skills are stilted; traditional views are holding back new methods of working. The industry remains slow and unresponsive. As the human population continues to expand, with now over 7.5 billion people on the planet, we need more suitable sustainable built environments and homes. Even in the developed world, the demand for housing outstrips supply, largely due to the resource intensive traditional methods that we continue to use. Anthony Sully offers valuable thoughts based on a vast body of experience that provide a new perspective. The insights allow readers to be more considerate of lessons learned, equipping professionals with foresight to improve design and avoid repeating the same historical mistakes. Importantly, the book explores expectations of the home and the operational standards that the industry needs to meet. Through the review of off-site manufacturing, an informed position is gained on engineered building and what might be achieved with future prefabrication and an Eco Home standard.

Sully's lens places people and technology together, so that we are better able to understand housing and what can be done to elevate standards. Practices in the field of construction and design have largely been pedestrian. A reluctance to change has created problems with practice, exemplified by the catastrophic events of Grenfell in the UK. A relaxation in standards and lack of consideration for design has seen buildings incorrectly designed with inappropriate thermal insulation, inadequate acoustic performance and problems of function and accessibility. Fundamentally, buildings should be properly designed, being considerate of the social and physical requirements and properly integrated within the built and natural environment.

Buildings are complex systems that need to be treated as whole systems, with performance standards that need to be met and maintained. Performance should exceed and not just meet the expectations of the building users. The autonomous home may be just around the corner, but without understanding the needs of the occupant and function of the building, within an integrated environment, the potential of comfortable living will be lost. The industry can no longer be content with its current mode of delivery.

Few challenge the conservative approach to building, but those who do can provide new and innovative methods of building. Master plans for energy efficient, energy flexible developments, adopting a holistic approach to living and working are becoming a reality, even if only on a prototype scale. More sustainable villages and cities are emerging; however, different approaches are required. It is good to see that Anthony Sully's work addresses the complexity of the home exposing some of the mistakes that designers have been making, by adopting traditional practice.

Many of the ideas are not new, but few have positioned them together providing an alternative way of approaching building. Not all ideas will be readily adopted or suit the palate of the technologist, developer or designer, but this publication will challenge views. The focus on design and technology will ultimately encourage all who operate in construction to reflect on what they do and the standards they want to achieve.

Leeds, UK

Professor Christopher Gorse
Sustainability Institute
Beckett University

Preface

This book is carrying out certain functions such as reporting, conveying a message, analysing, observing, appraising, informing and proposing. My writing has been stimulated by current events, inspired by various observational sources and engineered by my own desire as a designer to help improve an overwhelming socially disastrous situation and an architectural moribund wilderness when it comes to the mass estate housing sector.

Alain de Botton[1] made some films on the state of housing in the UK in 2014, and he commented that whilst the design of these houses had no design merit or personality (apart from the wretched pastiche of past styles as in Poundbury, Dorset), they seemed to have affected the occupants to the extent of reducing their will to be neighbourly or socialise, which results in them staying indoors watching TV. Rather a sad but relevant observation no less. When he questioned a builder about the pastiche architecture of one of his developments, the builder replied 'We just give the people what they want'. Now this cannot be true because the public may express opinions but they are derivative and are not the result of professional research. It is part of the builder's sales pitch but it is delusional. The public cannot be blamed for poor house design. It is up to the professionals connected with the housebuilding industry to provide a better service. Throughout my own research, I have been disappointed that I have not found any serious attempt by anyone to re-examine the process of designing a house rather like designing a product or as a mechanical object that has a shell, entrance and exit points for people, light, services and contents. The whole sector is governed by out-of-date processes and institutions that are stuck in traditional methods and misguided goals. There has been much progress in building technology and services, and the enveloping awareness of the importance of sustainability and energy conservation, but this has not been matched by the actual design of the spaces that define a house.

[1] Alain de Botton, born in Switzerland in 1969. Writer and philosopher dealing with subjects such as travel, architecture and literature.

Dr. Peg Rawes[2] also made a film recently with Beth Lord[3] entitled 'Equal by Design' but it tended to focus more on affordable housing. She and her colleagues summarised how the housing crisis was impacting upon the environment and the well-being of people but it did not get to grips with the fundamental inactivity of investigative house design research. In the promotion of well-being, it came out with platitudes for people that they should not aim to be happy, nor should they be pre-occupied with money, but rather they should look after themselves. This guidance is designed to help them reduce tensions by not setting impossible goals for living. I found this rather belittling and patronising although it is meant to be well-meaning.

The Design Commission's recent report[4] 'People and Places' states what should be done but it is yet another example of avoiding the 'how':

Government should establish a formal cost-benefit analysis of how design elements impact on behaviour in the built environment.

Housebuilders should be compelled to employ appropriately qualified urban designers, architects and landscape architects to ensure that schemes meet the highest standards of housing and neighbourhood design.

The authors highlighted 'the need to intervene on both design and social features of residential areas to promote mental well-being'.

What are these 'design elements'? I just wish someone would pay more attention to how families live in modern life and accept that the growing phases of children need more serious attention. Sometimes throughout the writing of this book, I have felt like a detective trying to solve a crime and crime it surely is. To witness the building of thousands of estate houses in the past and the present day, knowing that the vast majority of them will have to be demolished for not being fit for purpose is a tragic situation. Buying a car is the next big purchase in a person's life and, like many products, if fails to satisfy then it can be sold or exchanged quite easily. Not so a house. People can be more easily locked in financially or geographically and can end up spending years of misery. It has also been disappointing, but not surprising, that many companies whom I have approached have refused to participate in my research for this book. Their unwillingness could suggest that they have something to hide. Here are some of their replies published anonymously:

[2]Professor Peg Rawes, Programme Director M.A. Architectural History, Professor in Architecture and Philosophy, The Bartlett School of Architecture/UCL.

[3]Dr. Beth Lord, Reader in Philosophy, Head of Philosophy and School Director of Postgraduate Studies, School of Divinity, History, and Philosophy, University of Aberdeen.

[4]March 2017. This report was written by Jack Tindale (Manager, Design, Policy Connect) and Naomi Turner (former Head of Manufacturing, Design & Innovation, Policy Connect).

Thanks for your email. Unfortunately this is not something we would be interested in taking forward. We wish you success with your venture.

While we have been thrilled by the level of interest in our facility, we are wholly focused on completion and operation of our facility so regrettably, we will be unable to engage with you regarding your book.

Further to your recent enquiry, I have now had the chance to speak to our Directors and unfortunately in this instance we are not prepared to give our permission for the use of our photographs/plans in your new book.

I have spoken to the relevant people at the factory and I'm afraid they are extremely busy establishing our products and are not in a position at the moment to engage with this kind of research.

My obvious thanks go out to all those individuals, companies and organisations who have responded positively to my quest by making contributions of advice, suggestions and allowing their work to be published. I am well aware that this whole field is highly complex and has evolved through centuries paying witness to transitional phases of community development and political upheavals. I enter this arena as an experienced house owner and family man, and therefore, through my own career of designing and teaching, I have become aware of how much this housing sector has not provided the homes that people expect or need. Something has to be done.

Cambridgeshire, UK Anthony Sully

Acknowledgements

I would like to thank the following for assisting me with my enquiries and providing me with feedback to my design proposals:

Rob Delius of Stride Treglown Architects.
Oliver Grimshaw of Hanse Haus.
Adam Morton of the National Housing Federation.
Professor Chris Gorse of Leeds Beckett University.
Harvey Fremlin of The National Self Build & Renovation Centre.
John Kellett, Architect.
Reuben Wilkinson of Touchwood Homes.
Svenske Homes.

Whilst every effort has been made to seek approval for the use of information and images from contributors, I apologise in advance for any inaccuracies that exist.

Contents

List of Figures

Introduction

Housing is Britain's top policy issue. It is the "crisis" of our day. The Guardian, 1 October 2015

And they all get put in boxes, little boxes all the same

There's a green one, and a pink one

And a blue one and a yellow one

And they're all made out of ticky tacky

And they all look just the same.

'Little Boxes' is a song written and composed by Malvina Reynolds in 1962, which became a hit for her friend Pete Seeger in 1963, mocking suburban housing in the USA.

There is a housing crisis in the UK, which amounts to a shortage of well-designed dwellings in the mass estate[5] homes market. Government agencies and local authorities are planning to address this problem by encouraging building developers to build thousands of homes all over the country from single dwellings to multi-occupied dwellings serving a broad section of the community. I am not dealing with 'affordable or social housing' or housing for the homeless as this is beyond the scope of this book. This shortage of housing stock is as a result of population growth due to people living longer, the birth rate and immigration. According to the Office for National Statistics (GB), the population in the 1960s for the UK was 52,372,500 and by the 2010s it increased by 10 million. This places a tremendous burden on such a tiny island. These dwellings are built usually as a mixed development estate of up to a few hundred with an identifiable style. Amongst these is the 'estate house', which is the focus of this book. Despite these good intentions, the whole situation is plagued by the dominance of about ten major housebuilders, which the government acknowledges in the following:

[5]Estate: traditionally described a large country house and land of a wealthy owner. This meaning refers to the masses of housing estates built for the majority of the population.

Fig. 1 Typical modern estate grouping in UK. *Photo* Author

In a separate report by the Communities and Local Government Committee, MPs urged the break-up of the dominance of big housebuilders in order to create more competition to "fix the broken market". It said: "If we remain overly reliant on a part of the industry that has little incentive to change the way it works, then the country will not be able to deliver the new homes it needs." The Telegraph. 29.4.17

Also:

And while demanding the demolition of hundreds of council estates to free up land for redevelopment, the nine largest building companies currently sit on land with planning permission to build 404,040 new homes in the UK – further driving up the price of land and the housing they build on it. Woodberry Down, Architects for Social Housing,[6] May 2017

These housebuilders have been accused of sitting on 'Landbanks' which is land bought at a time when prices were much lower than they are now. Years later, they build on this land at high densities and sell the houses plus land that has escalated in value making them a huge profit.

The top ten housebuilders, which together have a market value of £37.5 billion, control almost half the market—up from 9% in 1960. Another trend that is contributing to this shortage is foreign investors developing and buying land and property, which is sometimes left empty whilst they sit on a growth product. As the Guardian published:

[6] *Class War on Woodberry Down: A National Strategy.* Published by Architects Social Housing May 2017.

> *It found "93% of flats in one of Manchester's biggest housing developments have been bought by foreign residents or companies registered overseas. Only 17 of the 282 flats were bought by British residents and only two are being lived in by UK owners."* Guardian 8.4.17

Housebuilders vary in type and are selected according to their building capabilities and skills, their contractual and management skills, and their ability to compete on price on delivery of these mass products. There is also the process whereby a buyer can organise the design and build of their own home through Self Build or Custom Build working with a developer. Chapter 6 defines the range of structures and materials that different builders will specialise in, such as brick and block, concrete, steel frame, timber frame. These builders are also categorised in terms of their scale of production, their preferred methods of prefabrication, and how far they can reach the standards of Eco Home status. It is interesting to note that architects are mentioned in most of my case studies, but when it comes to the large developers such as Barratts and Wimpeys, architects do not seem to get a mention. This is because they are in-house and have less status than the independent practices. Clive Aslet, former editor of Country Life magazine, wrote:

> *Volume housebuilders regard architects with contempt, don't employ them and we all have to live with the horrible results.* The Times, 30th November 2017

Another observation:

> *I cannot disagree that housing design needs a radical overhaul, however, the major problem currently is that the main players are not really interested in either building more houses or improving designs either to make them more affordable or more eco-friendly, despite what all the politicians say a shortage always keeps the market high. Also the major housebuilders have no incentive to build out sites rather to build as sold, this again keeps the prices high and denies any real benefit in using modern methods of construction. They also claim that their Clients want choice in fixtures and fittings so they cannot build ahead or pre-empt the finishes.* Mike Duhig, Modulogic Ltd.

It is absolutely no use building thousands of new homes if they do not provide sufficient space and storage so that people can lead proper decent lives, and that they continue to ignore the existence of children and their needs. The fact that current house design displays a paucity of imagination as well, by repeating well-worn, conservative and established plans and room descriptions, confirms the need for a total shake-up with the whole sector. Since 2004, the average house floor area has shrunk by 4 m^2 according to the RIBA. I am encouraged by a debate that was led by Jo Churchill MP in the House of Commons on the 13/12/17 when she said:

> *One's home is the biggest purchase that many of us will make in our lives. The fact that there is so little consumer protection attached to the purchase of new homes needs addressing. It is staggering that one is better protected when purchasing a kettle than when buying a house, given that the average house price in October was £223,000 and the average price of a kettle is £25……… The latest national new home customer satisfaction survey showed customers' dissatisfaction had risen to some 98%. Not all people are dissatisfied with their homes, but that shows that an alarmingly large number of people move into their new home, full of expectation, but are left unhappy with the quality therein.*

At the time of writing, the government revised the National Planning Policy Framework to assist with the housing crisis and have produced a 70-page consultative document. The only mention of design is in the following recommendation:

e) *the design is of exceptional quality, in that it: - is truly outstanding or innovative, reflecting the highest standards in architecture, and would help to raise standards of design more generally in rural areas; and - would significantly enhance its immediate setting, and be sensitive to the defining characteristics of the local area.*

This of course is typical of such a document that states 'what' but never 'how' which is what the public want. My argument is that there is no innovative thinking in house design and no signs of how it is to be achieved. It is not about style or technology of building but rather it is about re-examining what a house is in this modern world, what is its function, and how to design it as a product. The housing sector also does not acknowledge changes in modern living to such an extent that people are severely restricted in their own development and growth and, by that denial, have to conform to the wretchedly traditional housing format. The effect that interior spaces have on the well-being of people is completely overlooked in house design. Living in rectangular boxes has become commonplace to such an extent that people do not question them. They are not ideal, which is why I propose a more geometric approach to the design by investigating polygonal forms later in this book despite builders telling me that rectangular shapes are much cheaper to build than angular shapes. Well obviously! It is much quicker and cheaper to take a straight road to a destination than a more scenic route. We have a choice and we must budget for the better experience rather than the common and mundane. I have excluded disabled and wheelchair access users from my survey as they need to be especially catered for within purpose-built dwellings. All new housing developments according to current UK legislation should incorporate 10% of the houses as being equipped for disabled people. My aim is to prepare the conceptual design foundations for a prefabricated Eco Home type construction for mass production and take this to a level ready for further development. Time is money and materials are cost graded. But with faster factory production methods and faster on-site erection time, I fail to understand why we cannot produce an economic mass-produced prefabricated Eco Home (see Sect. 6.7) solution that is not a box and that satisfies the market. When it comes to clothing, there is a vast range of retail outlets that serve all income groups of the community. In other words, you can buy a cheap coat that is still warm and functional, but with housing it is becoming increasingly difficult to serve all levels of the public with a decent, spacious, sustainable and functional house. We are dealing with people's lives, not their leisure, not their workplace, not their healthcare, but where they live for two-thirds of their life.

I should mention that there is a lot of good work being undertaken by organisations like Studio Real[7] and the Built Environment Trust[8] at the Building Centre in London. The BET held an exhibition plus speakers entitled 'Making Wellbeing: from birth to death' in the built environment. Architects such as Sarah Wigglesworth, Peter Barber and Alex Ely explained some of the good work they are doing when it comes to planning and landscaping a neighbourhood and the supporting networks that should be provided to make life better for the residents. However, this does not dilute my mission for re-examining the interior provision of houses, which is sadly neglected and is what this book is about.

Chapter 1

Housing is a huge subject so I have decided to select and prepare examples from recent history. I will not be examining in-depth neighbourhoods or the reasons for change from one period to another but merely referring to the interior planning of selected houses. Nor will I compare costs of building as this is dependent upon land prices, location and the current economic climate.

This chapter takes a brief look at the evolution of the 'home' to set the present day in a historical context by examining various projects as Historical Case Studies. These begin with one house from the seventeenth century, one from the late nineteenth century before we descend upon the twentieth century. Whilst all the houses provide the basics for living, each house sequentially presents us with a new development that reflects changes in social habits or changes due to technology.

Chapter 2

This chapter examines recent examples of a variety of houses, apartments and furniture as case studies and through a comparative study exploring why UK estate housing is so bad and identifying particular weaknesses and strengths in the design and planning of the interior facilities provided. This survey is in response to the many cases of poor housing whereby people have been robbed of their dignity as the following confirms:

---

[7] Studio Real, Architects and Urban Planners, Oxford, UK.

[8] The Built Environment Trust exists to explore and encourage innovation in the built environment. It is an independent charitable organisation that provides support for educational, research and cultural activities. The organisation originated as The Building Centre, which opened in Central London in 1932.

The Guardian newspaper published in 2015 that the UK were in denial of Human Rights (HR)[9] when it came to housing. This illustrates how important a topic housing is and why I am writing this book. Usually, HR deals with social injustices in law and politics so it is even more alarming to discover housing has climbed into the HR bed of worthy causes.

> *The UK is in breach of its own United Nations human rights commitment to provide people with adequate homes because the housing crisis is so serious, a consortium of leading housing charities has warned. They cite soaring housing rental costs, unhealthy conditions in homes, and rising levels of homelessness and warn of "profound issues of lack of supply, increasing housing costs, lack of security of tenure and homes of such poor quality that they are unfit for habitation………… Since the 1980s we have lost any concept of housing's social function, and that is why protest movements are gaining ground. Without decent housing, you can't experience an adequate life in society, but now housing is seen just as an asset."* The Guardian, 28 April 2015

Chapter 3

The main intention of the book is to emphasise how modern living and family life are being ignored in favour of well-worn past precedents, and that it is time to explore ways of accommodating the family and children in a design proposal that is more about being responsive to human need. In this chapter, I am using all my research to inform on a design approach that I believe addresses the problems there are in the current designs of estate houses. I identify those workable areas that are familiar to a family home. An analysis of these activities, which suggest space and storage required, plus an understanding of daily family life informs the basis of planning. Due to progress in technology and pressure on space, modern housing no longer provides a fireplace (wood-burning stoves sometimes creep in), a larder, and in many cases a utility room as the washing machine and drier are usually incorporated into the kitchen. Older houses just had one bathroom but now an en-suite bathroom is added to the family bathroom as a permanent feature. Paul McGrath calls upon the RIBA to lobby the government on improving the quality of new houses:

> *If one of the most important challenges of modern times is satisfying demand for decent housing, it should matter to architects how the RIBA engages with the government on such issues. …………More importantly, across the board there is an almost total lack of understanding of how people live, or want to live, in various housing types which is arguably more important to understand.* Paul McGrath[10]

---

[9]The Human Rights Act 1998. The Human Rights Act 1998 sets out the fundamental rights and freedoms that everyone in the UK is entitled to.

[10]Paul McGrath, *Why space standards are a bad thing for the housing crisis.* BD Online 6 Jan 2015.

Chapter 4

This begins with a geometric analysis using a cellular gridded planning basis identifying what space is required for one person to perform certain activities. I then proceed to use the activity analysis from Chap. 3 activities in order to decide on space allocation and their relationships.

The geometric preference to pursue was polygonal, and I explain why this is a more human and organic route than the monopolisation of the rectangular grid that is all so familiar. I show some historical examples and some houses from America that features the octagon in planning. The octagon emerges as the preferred shape to apply to the planning of my proposed design. My research also reveals that anything made in square form is bound to be cheaper than an octagonal form simply because the octagon's $8 \times 135°$ angles create more energy and complexity in their connections. This proves to be my biggest battle with builders and developers who are motivated by the most economic solution possible. The preference for the safety and simplicity of a square is that it is tried and tested and it is called conditioning. People are simply not used to living in and using octagonal-shaped spaces. We have all grown up in boxes. So my mission in Chap. 5 is to prove the viability of chamfered spaces.

Chapter 5

The design approach I have taken in this chapter is that the architecture grows from the interior outwards. The interior shape of space defines the 3D form of the building. Two Stages show the development of the design process highly influenced by the conclusion that an octagon should form the basis of planning. This also illustrates the importance of reducing the impact of the square, which is so empty, common and moribund in a prison-like sort of way compared with the spiritually enlightening octagon. Stage 1 follows octagonal cellular relationships quite faithfully. Stage 2 combines two cell types to eventually produce the final more rationalised and workable design proposal that can form the basis of a 2, 3, 4 and 5 bedroomed house being built with the same floor plate without building any extensions. There are three different roof options for the three-bedroom house emanating from the same plan. A sample layout of an estate of 12 houses is shown combined with a study of varied types of access to the rear garden. The last section argues the case for the standard hinged door to be replaced by the sliding pocket door. I should warn readers that as I am changing the perception of what a home should be, I am also taking the opportunity of reviewing the names of the rooms and spaces that we are so used to using.

Chapter 6

This chapter summarises leading methods of prefabricated construction and services that are dominating the housing market in terms of sustainability and energy conservation. I have relied upon many sources of reference in order to provide me with the appropriate technical data. I describe some early examples of prefabricated houses so we can see how far the industry had progressed. Ultimately, I am using the chapter and collaborators to present all the options available in order to arrive at an Eco Home standard of construction for my final version house design in Chap. 5.

Why Am I Writing This Book?—A Lack of Design Innovation in Housing

I have chosen to write this book because I believe that there is a crisis in mass house design as confirmed above that is due to current legislation, inflammatory property values, poor workmanship and myopic design vision where children have been almost totally ignored in their growing phases from birth. This has prevented real innovation from taking place, and hence, the public have been denied the service they need. They are being duped by clever and seductive sales marketing techniques such as landscaping and external architectural re-modelling that continue to conceal repeated plans and conservative designs that do not represent progress of any kind. I will therefore use this book to address such issues and hopefully promote some way forward. I repeat that I am NOT dealing with emergency housing, social housing, housing for the unemployed or affordable housing as these are extreme circumstances beyond the scope of this book. There are of course many laudable schemes that have been built to house such people but I am dealing with another kind of emergency, which is about restoring some human dignity to housing in general. There is currently an alarming increase in the UK of people suffering from mental disorders of various kinds, and one contributory factor can be the poor quality of their home. Family life can become stressful with the lack of privacy, and a lack of space to carry out certain activities, which often have to overlap and be shared with other activities (see Sect. 4.5.4).

There are two other important reasons for writing this book:

(1) **Architect's detachment**

.When seeking more information on the interiors of one particular development, an architect replied that they were not responsible for the interiors and handed that over to someone else so they could not help me. Yet they had shaped and planned those spaces with the supposed realised perception of how they would be used. This kind of response betrays the disconnected attitude some architects have between the external form of the building and the interiors being planned.

(2) **Housebuilder's unwillingness to be involved in this book**

Having summed up my book's purpose and collaborative nature, more than one housebuilder I have approached expressed a desire not to be involved or not to release any plans even though I had hinted that their work could be shown in the book. I can only assume that they feared some criticism from me and were unwilling to consider working on a new design that would be a departure from their standard work. It sort of reminded me of the head-in-the-sand conservative attitudes that prevailed in the UK since the early 1960s, which contributed to the collapse of our industrial heritage. Another replied to me about my splayed corners (see Sect. 4.5.3): '*I can't see any building technical performance benefits to be had from changing the corners other than they may suit your own eye*'. You will observe dear reader that such proposals as you may come across in this book are not spurred on by aesthetics but by rather more deeper reasons related to human well-being.

Julia Park, who is Head of Housing Research at Levitt Bernstein Architects, wrote the following:

> *It will always be hard to argue that a lack of space could ever be life-threatening, though plausible that it could be life limiting. Even so, most homes are only too small, if 'too many people' live in them and trying to control actual occupancy is almost impossible. People tend to buy as much space as they can afford in the place they choose to live. For the vast majority of households, that means at least one spare bedroom – and for many, that's the extra space that makes their home workable.* One Hundred Years of Housing Space Standards. Julia Park. January 2017

I find that this statement dilutes the urgency facing us and is a conservative meek acceptance of the present situation. Lack of space IS life-threatening (see Sect. 4.5.4). Homes are small not because of overcrowding, the reasons for which can be complicated, but rather because the occupants cannot live properly. Then to measure the workability of a home by having a spare room as though the added square metre will make it all right is ludicrous. The 'extra' space should be absorbed by providing each activity with proper space and storage to function. The common perception is that space reflects the economic status of people and that the inherited squalor of the nineteenth century was the starting point for measuring what families required rather than a re-evaluation of their total needs which I maintain is well overdue.

Today we are suffering from a social disintegration of the family structure, human inter-communication being replaced by media devices, real experiences being challenged by the digital world, and the high street individual shops facing extinction whilst competing against on-line buying and being replaced by estate agents and charity shops. So the home is not only an important refuge from these social pressures, but more importantly, it is a place that confirms the beginning of a family union, provides a safe haven for bringing up children, provides a base for going out to work, provides a platform for a child to become an adult and it can be a place where the creativity and persona of its occupants can flourish.

Conclusion

This book is concentrating on the design of a house for mass production that will provide proper storage, space and privacy that has been neglected in the past, although many of the case studies are of a variety of dwellings as well as furniture. Currently, people are not moving house so much in the UK, due to not only the rising cost of housing but also because they have more disposable income to spend adapting their existing house. Spending money on a house not only covers any structural alterations but also acknowledges the increase in gadgets, electronic goods and media devices that have become a powerful factor in people's domestic lives.

Unfortunately, the rise in property prices in recent years has affected first-time buyers by making it extremely difficult for them to get on to the property ladder. The market has also been inflated by rich foreigners buying up real estate in order to make huge profits. A company called Pocket Homes in London builds prefabricated one-bedroom apartments for London's young city workers. They are compact and priced at least 20% lower than the open market. They design homes that suit the lifestyles of busy working singles and couples, and they make them affordable by providing minimal facilities with good storage and a secure bike store.

> *Lucien Smithers, sales director of Pocket Homes, said 'Young people don't need much space since they no longer have the physical belongings of previous generations. The most valuable things they own are their phone, laptop and bicycle.'* The Guardian. April 16th 2016

I would like to add a comment based upon my own experiences of living in various communities. And this is my point: I would like to emphasise 'communities'. Whatever housing scheme is planned, it is vital that the residents can meet socially either within an existing network or via a new development of pub, shop, school, etc. The most dispiriting thing about many housing developments in the UK, both old and new, is that the majority of people just do not socialise as much as they used to a few decades ago (see Preface). Apart from the many clubs and societies that attract people with particular interests, the pub was the place of voluntary associations: The place where people would freely interact with known regulars, or invite discussion with strangers talking about everything under the sun. Nowadays, there are many forces that have helped this disassociation: pub closures all over the country; the breathalyser and the smoking ban; more home entertainment combined with cheap supermarket booze (see Sect. 3.2.6). Of course over the centuries, the church was the main focus for social interaction, but as religion is on the wane so are the places of worship.

This book will ignore location as the study is designed to investigate improved approaches to suitable prefabricated housing for estate house design. The plan below is typical of many estates whereby the entrance road is more often than not the same as the exit road. Current legislation stipulates that 1 ha should have 30 dwellings of mixed occupancy. Many of such developments are tacked on to our towns and villages expecting them to absorb the impact of increased population and traffic, as well as making extra demands on local schools and other facilities. There are many understandable protests about such developments as the gains tend to be for the developers

Fig. 2 Typical housing estate plan—1600 houses near Braintree, Essex. *Courtesy* Acorn Housing

and political movers rather than for the residents. In this case, this plan is only for outline planning with no detailed agreed house design nor contracts yet made with builders. ACORN Housing, the developer, have told me that their aim is to address the poor standard of estate housing in the UK by ensuring that their houses are of good quality (Fig. 2).

Who Is This Book for?

It is primarily aimed at the professional grouping of developers, planners, architects, builders and building products' manufacturers, local authorities, as well as the educational grouping of students studying architecture, architectural technology, landscape design, product design, interior design and furniture design.

Hopefully this book will present an exciting enough design proposal in Chap. 5 to people wishing to help take it forwards. To quote Jo Churchill again:

> *Unless there is a challenge to the system to ensure that quality standards are driven up, there is little encouragement for those housebuilders who produce a poor quality product to raise their game.*

Chapter 1
Housing Background and History

A history of the evolving design of our houses offers a most revealing, if somewhat idiosyncratic, history of the nation. This is particularly the case when it's the humble home that is the focus, because its story makes clear how the tastes of the majority changed and how our domestic priorities—and aspirations—focused more on comfort, convenience and privacy.

Dan Cruickshank. The Times May 21st 2016.

Abstract The purpose of this chapter is to provide a backdrop to the estate house by conducting a brief historical survey of a selection of individual houses that help us to understand the evolution of the house, especially with regard to the provision of their interior facilities. I examine what makes a home in order to assist in my appraisals of both the houses in this chapter and Chap. 2. Whilst many of the accepted activities continue through time, it becomes apparent that their location can change or their descriptions change to become something else. It will be seen that houses had more space than those built now, but they had less daylight due to modern fenestration techniques.

1.1 From the Beginning

People have lived in dwellings since the beginning of time. Initially, they have found existing abodes or shelters within our landscape, before moving on to build their own abodes. Eventually, some people became such specialists in building that they could offer their services by building for others, either as a support for the community, or as a means of generating income whilst providing shelter for others. These homes began as unsophisticated amorphous dwellings, until societies developed with social hierarchies that needed other kinds of buildings to officiate, protect and rule the communities they served. The history of housing throughout the world is an enormous subject and far beyond the scope of this book. So I shall try and extract some key designs and references from the late nineteenth century, which I think will provide a platform for this book.

A. Sully, *The Estate House Re-designed*,
https://doi.org/10.1007/978-3-319-90397-2_1

According to Charles O'Brien in his book 'Houses'[1] England has over 22 million dwellings, of which one-fifth were built before the end of the First World War. This stresses how the twentieth century experienced a huge demand for housing due to population increasing, pressure on planning the infrastructure to cope with such an increase, and witnessed a huge variety of social pressures in supplying a wide variety of dwellings.

> *The model itself is the result of the collaboration of many people over many generations as well as the collaboration between makers and users of buildings and other artifacts, which is what is meant by the term* ***traditional****. Since knowledge of the model is shared by all, there is no need for drawings or designers. A house is meant to be like all the well built houses in a given area. ……………… The aesthetic quality is not specially treated for each house – it is traditional and handed down through the generations. Tradition has the force of law honored by everyone through collective assent. It is thus accepted and obeyed, since respect for tradition gives collective control, which acts as a discipline. Amos Rapoport, house, form and culture.*[2]

Rapoport reminds us how important tradition is in our lives in that it provides connections with our past and confirms our cultural roots. Of course herein lies the problem for designers as to how much tradition competes with modern ideals in the design of houses. Factors such as climate, landscape, socio-cultural aspects and religion influenced the housing form, construction and community layouts. People working on the land meant that dwellings were built to provide work, storage and living spaces. Eventually, the dwelling began to offer the basics of living, which was to do with working, storage, food preparation and eating, washing, and general living. The building had to be safe from attack, offer privacy and be fit for child rearing. The majority of the working population lived in appalling slum conditions with poor sanitation that people suffered from many diseases such as tuberculosis. It was not until 1875, that Joseph Bazalgette's[3] renowned sewerage system had been completed, improving the health of Londoners, and the 1878 Building Act established new bye-laws for housing development. It is quite remarkable and reversible that many of these surviving ancient dwellings have now become very desirable homes for the wealthy when converted to house a modern family's needs. Rapoport goes on to say

All housing needs to achieve four objectives in order to be successful:

1. *It needs to be socially and culturally valid.*
2. *It should be sufficiently economical to ensure that the greatest number can afford it.*
3. *It should ensure the maintenance of the health of the occupants.*
4. *There should be a minimum of maintenance over the life of the building.*

---

[1]*Houses, an Architectural Guide.* Charles O'Brien. London, Yale University Press. 2016.

[2]House form and culture. Amos Rapoport. New Jersey, Prentice-Hall, 1969.

[3]Sir **Joseph** William Bazalgette, 1819–1991, was an English civil engineer. As chief engineer of London's Metropolitan Board of Works, his major achievement was the creation of a sewer network for central London.

These are instructions to designers to acknowledge the time and location of a project that it should be affordable for the majority and should be designed to last. The progress of planning and designing these communities over the centuries has become very complex when considering the associated networks of industries and transportation. Nevertheless, one thing has remained very basic and constant in our lives, and that is the desire (not always achievable) to live with all modern conveniences, yet be able to sit outside to retain that contact with nature or whatever part of the outside world is accessible.

1.1.1 The Lifetime Homes Standard

This is a series of sixteen design criteria intended to make homes more easily adaptable for lifetime use at minimal cost. The concept was initially developed in 1991 by the Joseph Rowntree Foundation[4] and Habinteg Housing Association:

1. Parking (width or widening capability)
2. Approach to dwelling from parking (distance, gradients and widths)
3. Approach to all entrances
4. Entrances
5. Communal stairs and lifts
6. Internal doorways and hallways
7. Circulation space
8. Entrance level living space
9. Potential for entrance level bed space
10. Entrance level WC and shower drainage
11. WC and bathroom walls
12. Stairs and potential through-floor lift in dwellings
13. Potential for fitting of hoists and bedroom/ bathroom relationship
14. Bathrooms
15. Glazing and window handle
16. Location of service controls.

These were intended for generic housing and apartments of all types, which is beyond the brief I have set myself in this book. They seem to be straightforward and obvious to me.

Figure 1.1 shows my interpretation of a typical kind of drawing from memory that a child might do of a house. First of all, the image is realised by the front elevation because this is the strongest memory a child has of the importance of the entrance to his/her home. Secondly, the shape shows how important the roof is in

[4]The Joseph Rowntree Foundation (JRF) is a British social policy research and development charity that funds a UK-wide research and development programme. It seeks to understand the root causes of social problems, to identify ways of overcoming them, and to show how social needs can be met in practice.

Fig. 1.1 Child-like drawing of a house. *Drawn* Author

protecting the occupants from weather and human assault, and despite the many varieties of flat roofs that are built, the common perception in children is the standard pitched roof. The chimney is a fading element with the advance of central heating systems, yet it can still feature strongly in a child's mind simply because of the large quantity of older house types with chimneys that still exist, even if the chimney is redundant! Where is Father Christmas going to get into the house? The entrance door shows the way in and the windows are the 'eyes' of the house because that is how the house sees the rest of the world.

1.2 House Types

These house types represent a progressive development from early habitats, which was the creation of an upper floor for sleeping, away from the noisier general living areas on the ground floor. In Georgina Downey's book 'Domestic Interiors'[5] Trevor Keeble tells us that the notion of 'front door' and 'back door' emerged together with the hall as a neutral access so as to help with the division of the front room and the back room and upstairs and downstairs. In fact, James Wentling[6] points out that

[5]Georgina Downey, *Domestic Interiors.* Bloomsbury, London, 2013, p. 33.

[6]James Wentling, *Designing a Place called Home.* Switzerland, Springer. 2017, p. 74.

the 'back door' or 'side door' often became the common entrance for close friends or neighbours and even became the regular entrance for the residents because of returning to the house with goods that needed off loading in the kitchen. This seemed more efficient and direct than traipsing through the front hall. He also recommends that the back door be located near the garage, which ideally should have internal access from the house (see Fig. 5.15).

There is one more variation to the above which I explore in Sect. 5.3.2 and that is two detached houses joined to each other producing a 'Conjoined' pair of houses.

It is a variation on the semi-detached grouping, which began in the UK in the early nineteenth century by farm owners seeking a more efficient means of housing the labourers working on their land. The purpose was to combine two detached houses together sharing a dividing wall, which had the chimney stack as an effective means of sound division, but also harnessing the heat without loss through an external wall. The economy of means was creating by deception: two houses from one whole structure (Figs. 1.2 and 1.3).

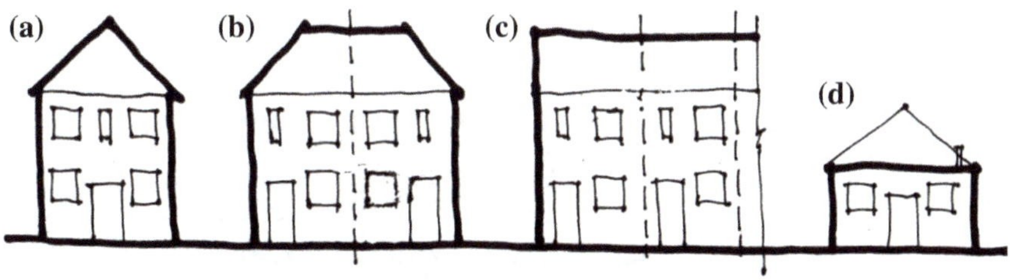

Fig. 1.2 **a** Detached, **b** semi-detached, **c** terraced, **d** bungalow/prefab

Fig. 1.3 Pattern book design for semis. Loudon 1846. More like conjoined. Not dissimilar to my own design in Fig. 5.30 but with garages. *Courtesy*, Pam Lofthouse

Figure 1.4 shows the more typical farmworkers' semi-detached cottages that were built. It has always seemed incredulous to me that someone would want to live in 'half a house', but they became so popular offering a price that people could afford that they became part of the British landscape. Fortunately, the two semi-detached case studies in Sect. 2.2 show enough design ingenuity as to make them as far removed from the traditional semi as possible. So my suggestion is that two houses that are joined together that appear to be two halves of a single house may be called 'semi-detached', and two joined houses that appear as two separate houses should be described as being 'conjoined'.

The bungalow, being essentially single storey, is perhaps the most derided of housing forms because hardly any have been built of architectural merit. It has a large roof area and hence much heat loss to cope with. It can occupy a large site with perhaps three or three bedrooms and is seen mainly as a retirement home. A staircase is a form of exercise, which would be good for older people, so the idea that decreased mobility demands restricted movement (no stairs) in the home is not helping. Bungalows are suitable for the physically handicapped but not for the healthily mobile.

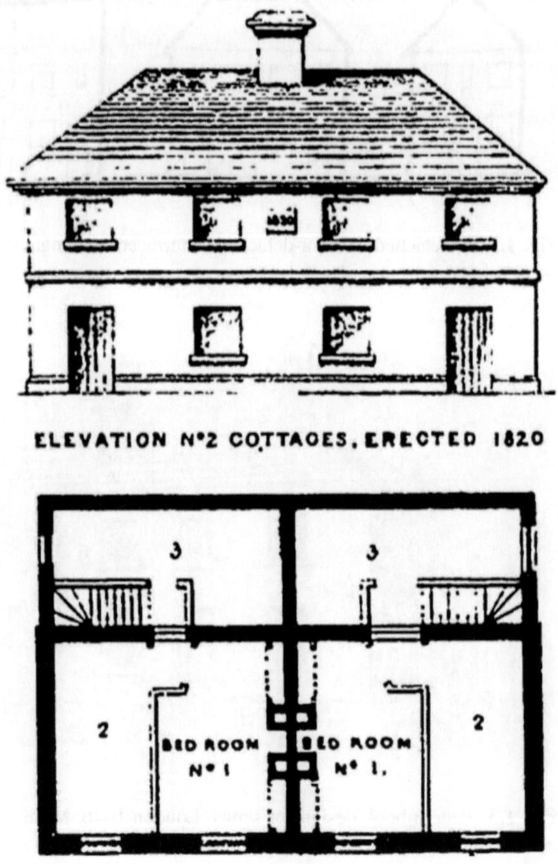

Fig. 1.4 Semi-detached Estate Cottages in Holkham, Norfolk, UK, 1842. *Courtesy* Pam Lofthouse

A house building can fall within subject headings such as 'domestic', 'residential', 'community', 'neighbourhood', social (or low cost) housing and so on. Many of the terraced houses in our towns and cities were built for the working population in the last 200 years and hence were a sort of mass-produced product because they were built in their thousands. The Victorian terrace, for example, is an extremely flexible form, offering a reasonably large space and the possibility of rearranging walls, allowing an occupier to stamp his or her identity on the accommodation. It is also flexible—it can be subdivided into flats or converted back into a family house. It is amazing to me that the grander Georgian and Victorian terraced houses still survive in the age of the motor car when they were never built with any garages, especially in town centres where a car park space is at a premium.

Figure 1.5 shows some town centre Victorian terraced houses with no garages and in a two-way road that is so narrow that cars have difficulty in passing each other with cars parked on both sides. This became so impossible that the council by painting white lines has allowed the cars to park half on the pavement to help relieve the situation. This of course is a totally unacceptable situation because it is dangerous for pedestrians. Yet a one-off house from Fig. 1.2 could also be built (using the same building methods) to a higher specification for a richer person and would subsequently have a higher grade of materials and decoration than the mass-produced ones. But strictly speaking, they were not 'mass-produced' in the sense that they rolled off the factory floor. However, there were many components, which were mass-produced from a factory, starting with the common brick then roof tiles, windows, doors and staircases. Smaller items in the interior were also mass-produced such as ceramic tiles, sanitary ware, ironmongery, kitchen fittings

Fig. 1.5 Typical Victorian terraced houses with no garages. *Photograph* Author

and fireplaces. All of these items, including the timber roof structures and partitions, were then assembled and fitted with great skill once the foundations, and main structure of the house was built on site. These homes would then be furnished by the occupants with loose items of furniture for physical support, soft furnishings that provided colour and pattern, and light fittings. The building itself was therefore an assemblage of products and finishes. The expanding middle classes of the late nineteenth century became very proud of their homes and their furnishing reflected upon their status. Locations can vary such as:

- Isolated rural
  Either farmworkers or local industry to wealthy commuters. Country roads.
- Village
  Community consisting of a mixture of local workers, shopkeepers to commuters. Church. Country roads. Primary school.
- Small town
  As village but larger population. Better road and transport system. Secondary school.
- Suburb—to a large town or city
  Not such a strong community as a town or village. Poor communication between people. Built for commuters. Could be boring with no heart due to its rapid expansion to provide housing on a mass scale for the workers. Good transport links.
- Inner city
  The working hub of a province with great road, rail and bus links. Housing for the wealthy and middle classes with the working classes being forced out due to increase costs in living. Transport important infrastructure. Places of leisure and entertainment, ceremony, education, law and administrative centre (Fig. 1.6).

Fig. 1.6 Typical Georgian terraced houses with no garages

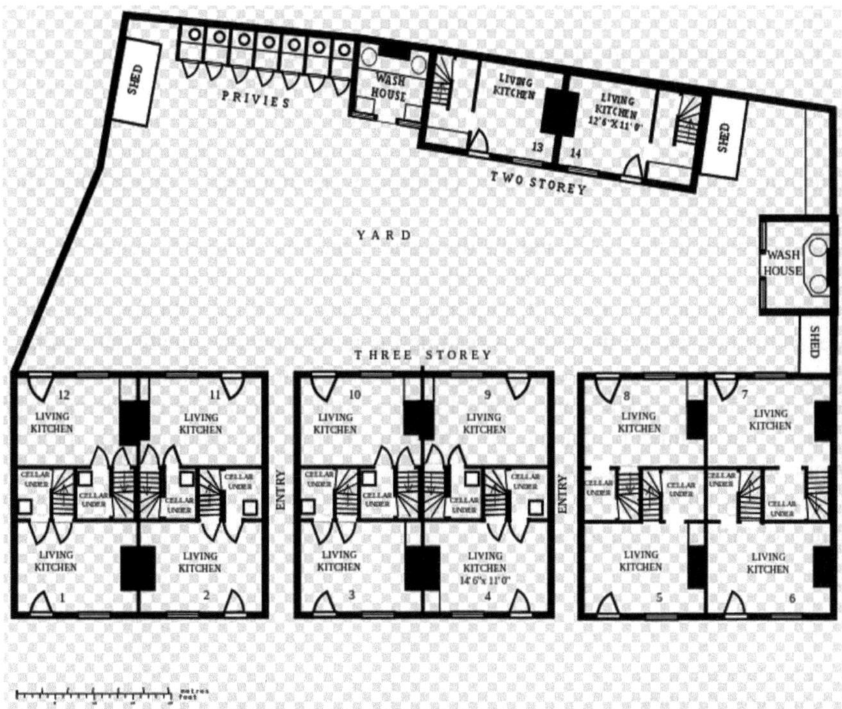

Fig. 1.7 Back to back housing plan. Wikipedia. Author Clem Rutter

There is one other house type that was built in the late eighteenth century in the UK for the poor, namely the 'Back to back house', a form of terraced housing. They were built for the rapidly increasing population of Britain's expanding factory towns. But they were soon denounced as being unsatisfactory, and the passing of the Public Health Act 1875 meant that no more were built; instead, bye-law terraced houses took their place. They were planned as a one room footprint of three or four floors for basic living, a recipe for slums if there ever was one with poor ventilation and cramped living conditions. Figure 1.7 plan shows these built in groups of four back to back with chimney stacks, winding staircases and a communal yard.

1.3 What Makes a Home?

A house is one kind of building in which people live and within which a person's life can begin. It is basically providing shelter and protection. If a family is living there the place becomes a home. It becomes a place where memories are stored, where time is measured by family growth, occasions, early schooling and a locking into the local community.

The charming scene in Fig. 1.8 is of a bygone age when the mother is sewing by the fireplace and the child deep in thought, writing with a quill pen on what appears to be a coffee table. To get the child to sit comfortably at the table a stool (piano?) is provided, but because her legs are too short a footstool is placed for her support. With toys on the floor, this space is almost dominated by the child's activities indicating that even in those days children were not really catered for (See Introduction) (Fig. 1.9).

Fast forward to the 1940s and the domestic scene has changed to acknowledge a combined lounge/dining room although still without a television. The family are pursuing the usual activities with a bit more intensity identified with their respective roles in the household: the father is reading the newspaper presumably after a hard day's work; the mother is sewing thus continuing with her well-defined duties as listed in Sect. 3.1, and the son is doing his homework. I do like the care being given ensuring that all three people have a lamp to work from instead of relying upon the ubiquitous central ceiling pendant.

The 1930s kitchen in Fig. 1.10 shows how much we have progressed in terms of quality of fittings, hygiene and design.

Fig. 1.8 Domestic scene from the book 'Childhood' written by Githa Sowerby and illustrated by Millicent Sowerby. Published by Chatto and Windus 1907

Fig. 1.9 Lounge/dining scene in 1940s. From 'The Housewife's Book' published by the Daily Express. *Courtesy* The publisher

Fig. 1.10 A small 1930s kitchen with cooker, fridge and sink. From 'The Housewife's Book'

In my lifetime, I have lived in bedsitters, flats, semi-detached houses, detached houses and odd conversions such as part of a large country house divided up into 12 residences, a chapel into a single residence, a farm made into four residences, and a granary divided up into four residences. In addition, I have visited a great variety of other abodes such as a narrow boat, a converted pub and many others. In every case, I and other owners have attempted to identify ourselves with the interior through a choice of furniture, fittings and decoration. We have put our mark upon the property. Interior decoration before the 1950s enabled people to strongly identify themselves with their surroundings because their visual qualities were looked upon with greater attachment compared with people today. Before TV, computers and the radio, the interior environment was designed to hold our gaze and attention with the intricacies of the decoration whilst immersed in conversation, whereas today there are many more devices calling for our stronger interactive participation over the interior furnishings, all at the expense of conversation!

I shall attempt to select examples of houses through recent history simply to show the main planning ingredients that make up their character and how the spaces are used. This is not an exhaustive survey of decoration, nor an account of styles of living and influences of technology and social trends, all of which are well covered in many other books on the subject. This is rather a summary that enables me to trace a thread that will support my conclusive efforts.

1.3.1 Historical Legacy—Audley End

You could say that the great English Country Houses from the Middle-ages onwards represent the highest ideal of a home that matches the status, position, power and wealth of the owner. The bigger they were the higher the status and the closer they were to royalty (Fig. 1.11).

Figure 1.11 shows the front elevation of the three-storey Audley End House surrounded by 10,000 acres of gardens, lake and additional buildings. I have chosen this house because it is a prime example of an English country house that is still in scale with its visitors as opposed to being so large it would be beyond comprehension. Audley End was built on the foundations of a twelfth-century Benedictine monastery and is now only one-third of its original size, hosting some of the best Jacobean interiors in the UK. As with many country houses, the grounds were laid out by Capability Brown[7] and the interior details were supplied by designer/craftsmen such as Robert Adam,[8] with more recent classical redecorations by

[7]Lancelot Brown (1715–1783), more commonly known with the byname Capability Brown, was an English landscape architect.

[8]Robert Adam. (3 July 1728–1792) was a British neoclassical architect, interior designer and furniture designer.

Fig. 1.11 Audley End House. Early seventeenth-century country house outside Saffron Walden, Essex, England. Built by Thomas Howard, 1st Earl of Suffolk. *Drawn* Author

Dudley Poplack.[9] The whole existence rested upon presentation using the highest standards of craftsmanship of the day and in that sense a building of its time.

Figure 1.12 shows the plan of the ground floor and the whole building's internal floor area of house is approximately 525,000 ft^2 for four main bedrooms. The plans showrooms with a use that reflects how the house was maintained with servants and how the owners had their own designated quarters. The house was basically divided up into three groups of spaces: (1) servants' quarters, (2) owner's private quarters and (3) public spaces for entertaining. The whole lifestyle would be so far removed from the common population as to be from another planet. Pleasure is derived from the classical symmetry of the planning, as one space connects with another via decorative details and furnishings, and the progression through the house is like witnessing a theatrical production of various scenes. Views to the outside are heightened by projecting bay windows, which invite a place to rest and enjoy the view or partake in conversation. Contrast that with the situation today whereby the majority of these houses have had to be sold because the cost of upkeep and death duties has become an unbearable strain on the owners. Alternatively, they have diversified to create an income such as Woburn Abbey's Safari Park or the Palace House at Beaulieu which runs the National Motor Museum.

[9]Dudley Poplack (1930–2005) is a South African interior designer but settled in England and a fashionable decorator for the wealthy and royalty.

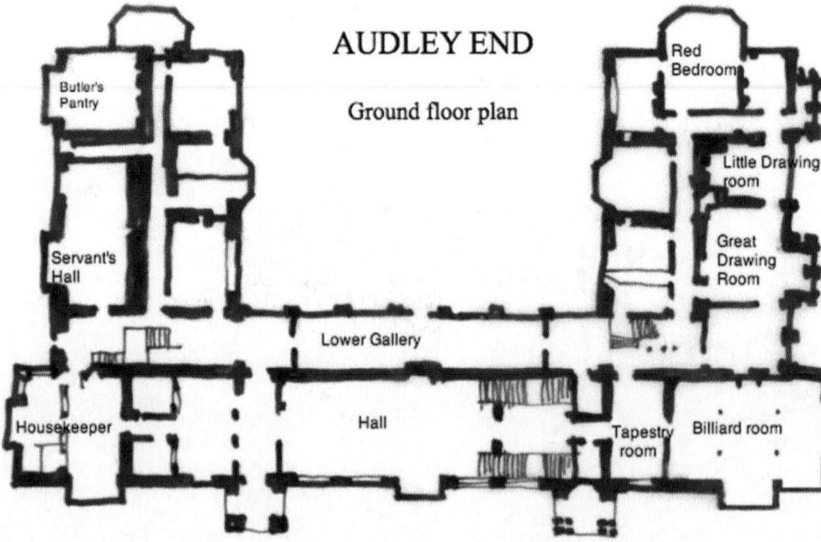

Fig. 1.12 Ground floor plan of Audley End House. *Drawn* Author

Some have been demolished, and the remainder has either gone to private buyers or has been left to the nation such as the National Trust.[10]

1.3.2 *Historical Legacy—Biltmore House*

The status of people in the modern world is very different and more varied than in the seventeenth century. The ruling classes have diminished in number and have given rise to people of corporate wealth such as bankers, financiers and property developers, not forgetting the incredible wealth achieved by professional footballers, and celebrities in the entertainment industries. The eighteenth and nineteenth centuries saw a rise in the great industrialists as the new wealthy group that earned their wealth as opposed to it being passed down through a titled family. One such excessive example is Biltmore House in Ashville, North Carolina, designed by Richard Morris Hunt[11] for the Vanderbilt family from 1888 to 1895, the largest

[10]The National Trust for Places of Historic Interest or Natural Beauty, known as the National Trust, is a conservation organisation in England, Wales and Northern Ireland, and the largest membership organisation in the UK. The trust describes itself as "a charity that works to preserve and protect historic places and spaces—for ever, for everyone".

[11]Richard Morris Hunt (1827–1895) was an American architect of the nineteenth century and a preeminent figure in the history of American architecture. Hunt founded both the American Institute of Architects and the Municipal Art Society.

house in the USA with grounds of 8000 acres. I have chosen this house because it is the product of a rich industrialist as opposed to a titled peer of the realm. Also, the house may be American, but the design is distinctly European in being very similar to French chateaux. The Vanderbilts, who were of Dutch origin, were at one time the wealthiest family in America, having made their money initially through shipping and railroad empires. With 250 rooms, it included 35 bedrooms, 43 bathrooms and 65 fireplaces all for one married couple! This represented the axiom 'biggest is best' but to an absurd level of excessive over indulgence to the point of repulsion, at least in the mind. But when you visit the house, as I did, you cannot but gasp in awe at the magnitude of everything. The act of living there for the family must have been daunting, for surely it was never a 'home' but rather a hotel. Unfortunately, it represents the worst of Victorian architecture in its borrowed themes from past styles. Frank Lloyd Wright,[12] one of the pioneers of the modern movement, had just come out of school when Biltmore was built. One wonders what he would have done if he had received such a commission (Figs. 1.13, 1.14 and 1.15).

The plans show how much the Americans loved octagonal shaped rooms, a shape to feature later in the book.

The whole spirit of the interiors was a direct follow-up from the splendour of the exterior and the landscaped gardens. The grandeur and formality were maintained through a processional sequence of small and large spaces, defined by a variety of plan shapes such as a square, a rectangle and variations of the octagon or part of. It was all about ceremony and dignity with lavish decoration and furnishing to complete an enclosure of mesmerising visual content.

Fig. 1.13 Biltmore House, Ashville, North Carolina. Designed by Richard Morris Hunt 1888. *Photograph* Author

[12]Frank Lloyd Wright (1867–1959) was an American architect, interior designer, writer and educator. Wright believed in designing structures that were in harmony with humanity and its environment, a philosophy he called organic architecture.

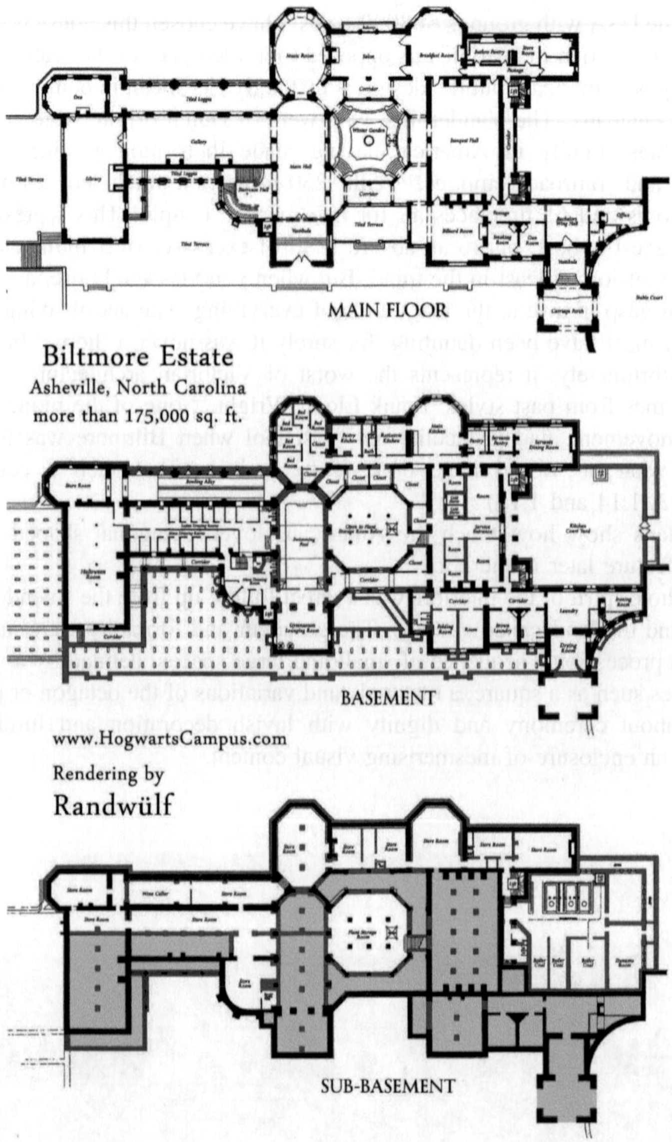

Fig. 1.14 Plans of Biltmore drawn by Randwülf. *Source* Wikipedia

Fig. 1.15 Pen sketch of Biltmore drawn by Author 1998

The Victorian period was a curious mixture of competing work ethics when it came to architecture and design. On the one hand, we have examples of revivalist architecture as shown above, and the mass housing demands to house a workforce from an ever-expanding unprecedented industrial growth as witnessed by the Great Exhibition of 1851.[13] And on the other hand, an Arts and Crafts movement focusing on craft and handmade goods spearheaded by John Ruskin[14] and William

[13]The Great Exhibition of the Works of Industry of All Nations or The Great Exhibition, sometimes referred to as the Crystal Palace Exhibition in reference to the temporary structure in which it was held, was an international exhibition that took place in Hyde Park, London, from 1 May to 11 October 1851.

[14]John Ruskin (1819–1900) was the leading English art critic of the Victorian era, also an art patron, draughtsman, watercolourist, a prominent social thinker and philanthropist. He wrote on subjects as varied as geology, architecture, myth, ornithology, literature, education, botany and political economy.

Morris,[15] which was a reaction against the increasing mechanisation of manufacturing processes. The Edwardian period was such a style that came out of the Arts and Crafts and Aesthetic movements. (C R Ashbee.[16])

1.4 The Aesthetic Movement

From 1800 to 1900, the UK population grew from about 11–32 million. The urban population increased from around 2 million to over 20 million. Since the industrial revolution in England beginning around 1760, there has always been a change in the design and planning of houses covering all sectors of society. City and town centres grew through the expansion of the money markets and as administrative centres dealing with the law and government. Country life changed through mechanisation of farming and the spread of other small industries. Even some companies built model villages for their workers such as Bournville[17] did in Birmingham. (See Sect. 1.3.2), or Port Sunlight in Cheshire. People were housed according to their wealth and status. These spanned from the working classes of factory and farm workers, to professional and white-collar workers, and finally the higher income group of owners and entrepreneurs. As this book is concentrating on housing for the masses, I shall focus upon the middle-income group house type.

1.4.1 Bedford Park—Arts and Crafts Movement Estate House

I have selected this house because of the quality of the design by one of England's most respected architects, Norman Shaw. The Bedford Park development by Jonathan Carr in Ealing, West London could be considered to be the beginning of the idealised garden suburb with tree-lined roads, welcoming and homely dwellings with gardens providing all the needs and facilities for workers returning from the smoke and grime of the city. Richard Norman Shaw[18] was appointed the estate

[15]William Morris, (1834–1896) was an English textile designer, poet, novelist, translator, and socialist activist. Associated with the British Arts and Crafts Movement, he was a major contributor to the revival of traditional British textile arts and methods of production.

[16]Charles Robert Ashbee (1863–1942) was an English architect and designer who was a prime mover of the Arts and Crafts movement that took its craft ethic from the works of John Ruskin and its co-operative structure from the socialism of William Morris.

[17]Bournville: best known for its connections with the Cadbury family and chocolate—including a dark chocolate bar branded *Bournville*.

[18]Richard Norman Shaw (1831–1912) was the most fertile, representative and immediately influential domestic architect of the late Victorian period in England.

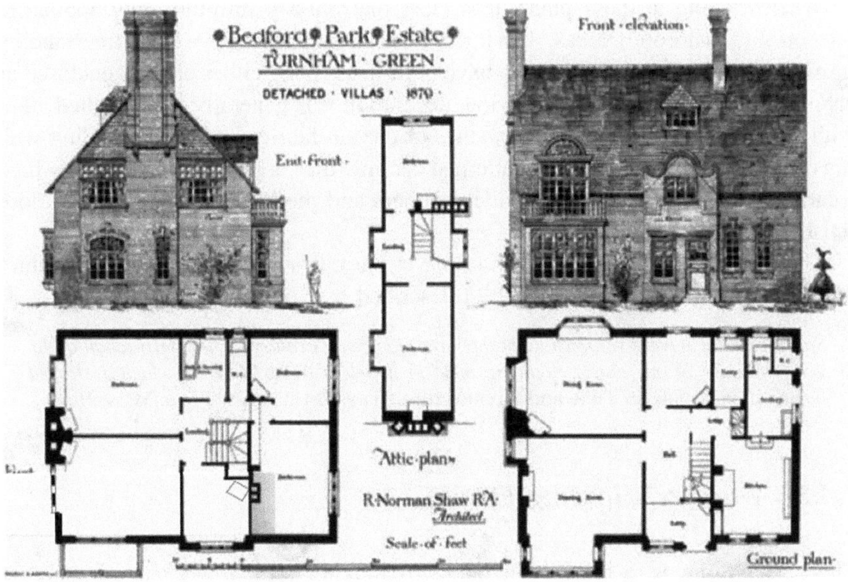

Fig. 1.16 Detached seven-bedroomed three-storey house, Bedford Park, Turnham Green, London. Designed by Richard Norman Shaw 1879

architect for Bedford Park in 1877, and he designed these houses in the Queen Anne Style.[19]

Other garden suburbs around London were Hampstead Garden Suburb, Letchworth (the first) and Welwyn Garden City. Ebenezer Howard proposed the Garden City concept in 1898.

The house shown in Fig. 1.16 is one of many that were individually designed with different 3D forms and layout using a combination of different materials. The seven bedrooms shared only one bathroom and a separate toilet on the first floor, and on the ground floor a living room, dining room, hall, lobby, kitchen, pantry, laundry, a spare room (study?) and another toilet accessed from the outside. You will note that fireplaces were in every room either along a wall or wedged into a corner because there was not any central heating at that time. The delightful ingredient in these types of houses was the modelling of form both outside and in the interior. The bay window sometimes with a window seat is the most welcoming aspect together with the ornate mantelpiece over the fireplace serving as a focal point. Two critical properties contribute to the atmosphere of the house: daylight entry and the warmth of a fire.

[19]The Queen Anne style in Britain refers to either the English Baroque architectural style approximately of the reign of Queen Anne (reigned 1702–1714), or a revived form that was popular in the last quarter of the nineteenth century and the early decades of the twentieth century.

When looking at these plans, it is clear that built-in furniture only applied to kitchen and bathroom areas, leaving all remaining rooms to have freestanding furniture. Library areas would also have built-in shelving either open or enclosed as cabinets. As with the Victorian period, decoration was generally a full-bodied affair with wall panelling, wallpapers, swathes of curtain fabrics and floor boarding with rugs. Hallway floors were often patterned ceramic tiles, a sensible introductory hard wearing surface between the outside weathers and the more delicate interior floor finishes.

The sad plight of much of the housing in our cities in the late ninetieth century due to massive redevelopment is best described by Sigfried Giedion[20]:

> *Speculators also were responsible in the later nineteenth century for the destruction of the very structure of our cities, engulfing them in hopeless living quarters which destroyed human dignity.* **Space, Time and Architecture**. Harvard University Press. Mass. 1962

1.4.2 Workers' Housing Estates

This selection helps to balance the books by looking at the lower income group of society.

Mention must be made of these, which either became known as social housing or council housing, and they were usually rented accommodation. The architecture though was very fine indeed even though they copied past styles. The interior provision was also to be commended but very basic.

From 1888—Port Sunlight the soap makers Lever Brothers, The Wirral, Merseyside.

From 1800s—Bourneville by Cadbury's the chocolate makers.

1915—Progress Housing Estate, Eltham, to house munitions workers from Woolwich Arsenal (Fig. 1.17).

The following characteristics define the special character of this Conservation Area:

- the area's period quality as a remarkable early example of Garden Suburb planning, with picturesque street layout, green spaces and relatively low density;
- the stylistic character of the houses built as English romantic cottages;
- the use of quality detailing and materials associated with fewer budget controls than normal, consequent on the need to build at speed in wartime;
- the product of talented architects in the Office of Works under Frank Baines;
- the consistent use of characterful originally oak front garden fences in a design particular to the estate;
- the use of natural mottled green and blue Westmoreland slate—often laid in diminishing courses;

[20]Sigfried Giedion. (1888–1968) was a Bohemian-born Swiss historian and critic of architecture.

Fig. 1.17 Typical semi-detached estate house on the Progress housing estate influenced by the Arts and Crafts era. *Photograph* Author

- the use of a limited range of front doors which differentiates the identity of the estate;
- the estate's particular picturesque cottage townscape style; note in Fig. 1.17 the 'jetty' overhang of the upper floor taken from the Tudor times.
- the survival of nearly the whole estate of 1200 housing units—with minimal infill or demolition;
- the completed character of the estate,
- the picturesque effect of designing in footpaths connecting the streets;
- the use of a standard six light vertical wooden casement window module.

Today's developers of housing estates would do well to study the above list accepting that modern technology will replace some of the above. What is impressive is the philanthropic nature of such developments as being of a worthy cause. Frank Baines, who was a former pupil of the Arts and Crafts Architect C.R. Ashbee was the chief architect for this development.

> *All local authorities wanted tenants to live in houses on garden estates where their lives would be improved to the benefit of the community and local industry. The between-wars designs started off trying to follow the "arts and crafts" style of cottage, but as schemes got bigger and funding became harder, the designs gradually became more functional and*

cheaper to build, but most authorities still tried to build for a minimum life of 60 years and this meant construction to a high standard. It also meant that the preferred building material was the well understood house brick, as the durability of concrete and steel was not yet proven. Surprisingly in the British climate, many of the houses built of timber have lasted well and those on the LCC garden estates look very good today. www.socialhousinghistory.uk

1.4.3 Space Standards

The Addison Act 1919 houses were usually three-bedroomed houses with lounge and scullery or lounge, dining room and scullery. Some had two, four or five bedrooms and generously sized back gardens intended for vegetable growing. At the best, they were built at 12 houses per acre. Importantly, they were built to the recommendations in the Tudor Walters Report commissioned by the government in 1917 (Fig. 1.18):

It is interesting to note the differences compared with today's expectations. There is no kitchen because the living room would have a cooking range and the scullery would have a copper boiler to heat the water, a bath and a gas cooker for occasional use. A separate bathroom would be upstairs. The parlour is a reception room similar to a modern living room's function. The total floor area including the

Tudor Walters Committee Recommendations

House with out a parlour	**Area sq ft (m^2)**	**Volume cu ft (m^3)**	**House with a parlour**	**Area sq ft (m^2)**	**Volume cu ft (m^3)**
			Parlour	120 (11)	960 (27)
Living Room	180 (17)	1,440 (41)	Living Room	180 (17)	1,440 (41)
Scullery	80 (7.4)	640 (18)	Scullery	80 (7.4)	640 (18)
Larder	24 (2.2)	-	Larder	24 (2.2)	-
Bedroom No. 1	150 (14)	1,200 (34)	Bedroom No. 1	160 (15)	1,280 (36)
Bedroom No. 2	100 (9.3)	800 (23)	Bedroom No. 2	120 (11)	960 (27)
Bedroom No. 3	65 (6.0)	520 (15)	Bedroom No. 3	110 (10)	880 (25)
Total	855 sq ft ($79.4 m^2$)			1,055 sq ft ($98.0 m^2$)	
Desirable Minimum sizes - Tudor Walters Committee					

Fig. 1.18 Table of recommended areas by the Tudor Walters Committee 1917

parlour and bathroom is 98 m^2. The scullery and larder have now disappeared to be replaced by the utility room, and the addition of a study is a reflection of home working demands.

1.5 1930s Housing—The Interwar Years

This was a period of suffering from the depression following the Great War with hardship being felt across all those countries that had participated in it. However, this seemed to instil a determination in people to create music such as jazz, and the arts that lifted people's spirits. Despite a turbulent economy and the National Debt in England, industry was recovering towards the end of the 1930s. The all-pervading styles were Art Deco[21] and Moderne leaving behind the wonderful Art Nouveau period. A whole range of historical styles became popular including 'Spanish Mission' and 'Cape Dutch', with their curly gables and neat brickwork emulating colonial architecture from the Americas and South Africa.

The country needed to build many houses to cope with the increasing population and suburbia expanded at a prolific rate. England's Central Government passed a series of Acts to cope with this problem such as: The Addison Act of 1919 and one in 1930. The 1935 Housing Act sought to attempt to regulate space standards which continued to be undersized as was the Parker Morris standards (see Sect. 1.5.6). Up to now, mass housing consisted of either row of terraced houses for factory and railway workers, or city terraces for the professional classes. The new demands were for housing the middle classes with their cars and their expectations of raising a family in pleasant tree-lined neighbourhoods, with parks and small shops that reflected village life. Some of these were called 'Cottage Estate Houses' such as the Becontree Estate in Dagenham, East London, (London County Council Architects, 1919–1935) and 'Wythenshaw' in Manchester (Barry Parker from 1931). These suburbs were planned on the edge of our big towns and cities coinciding with the spread of the London Underground network. The history of the London Underground began in the nineteenth century with the construction of the Metropolitan Railway, the world's first underground railway. From the beginning of the twentieth century, more lines were introduced and these became important travel arteries to serve the commuters working in central London. I shall first of all provide examples of special one-off houses before showing houses for the masses, so as to identify common needs.

[21]Art Deco, sometimes referred to as Deco, is a style of visual arts, architecture and design that first appeared in France just before World War I. Art Deco influenced the design of buildings, furniture, jewellery, fashion and many products. It took its name, short for *Arts Décoratifs*, from the Exposition Internationale des Arts Décoratifs et Industriels Modernes (International Exhibition of Modern Decorative and Industrial Arts) held in Paris in 1925. It combined modernist styles with fine craftsmanship and rich materials.

1.5.1 The First Modern House

This house is selected because it signalled the beginning of the modern movement.

The Modern Movement in architecture was a reaction against the revivalist styles of the nineteenth century and was seen as a cleansing of the soul compared with the folksy Arts and Crafts with all its embedded traditions. Architects like Frank Lloyd Wright, Mies van der Rohe[22] and the whole Bauhaus[23] school founded in Germany led the way in this design purification and the search for truth in the use of materials whilst embracing new technology.

High and Over Park in Amersham (on the metropolitan underground railway line), Buckinghamshire, was the location of the first modern house to be built in this country following on from the Bauhaus and the work of such architects as Le Corbusier.[24] It was designed for Professor Bernard Ashmole of London University in 1929 by Aymas Connell[25] of the practice Connell, Ward and Lucas, to take advantage of its hilltop position and the sun. It became known as the Sun House, and the Y-shaped plan took account of this with its three wings radiating out from a central hexagonal hall. It was divided into two properties in 1962 (Figs. 1.19 and 1.20).

Considering my interest in geometric cellular design later in the book, this hexagonal axial planning is a surprise to me for this period. Although the properties of the hexagon appear nowhere else in the design of the house except for the projecting staircase. What is interesting about this plan is the departure from the traditional rectangular with its entrance hall and front and rear rooms. The 'V' formation on entering is welcoming, and then one is greeted by a spacious hall with fountain. Sliding folding doors mean that they almost disappear, and they open to reveal the lounge to the left and a library to the right exactly the same size, both areas being 'reception' by nature. The third wing of the house opens up to the dining room yet another reception space. The kitchen is kept hidden contrary to modern trends of uniting the kitchen, dining and living spaces.

---

[22]Ludwig Mies van der Rohe (1886–1969) along with Le Corbusier, Walter Gropius and Frank Lloyd Wright, he is widely regarded as one of the pioneers of modernist architecture.

[23]Bauhaus was an art school in Germany that combined crafts and the fine arts and was famous for the approach to design that it publicised and taught. It operated from 1919 to 1933.

[24]Le Corbusier (1887–1965) was a Swiss-French architect, designer, painter, urban planner, writer and one of the pioneers of what is now called modern architecture.

[25]Amyas Douglas Connell (1901–1980) was a highly influential New Zealand architect of the mid-twentieth century. He achieved early and conspicuous success as a student, winning the British Prix de Rome in Architecture in 1926.

Fig. 1.19 High and Over Park, Amersham, Bucks, 1929. Architect Amyas Connell. *Courtesy* The Modern House

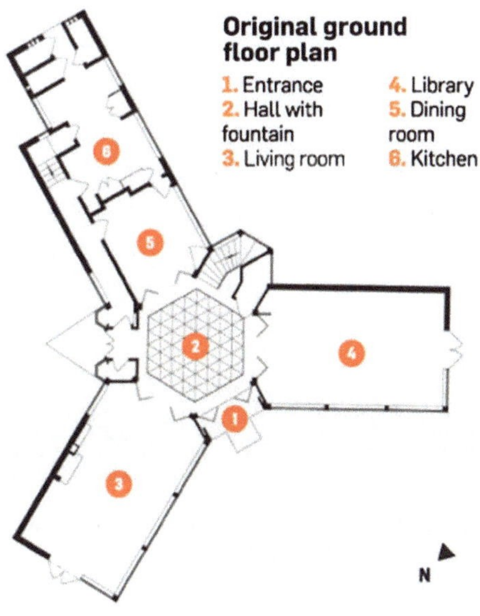

Fig. 1.20 Ground floor plan of High and Over Park

1.5.2 Hill House, St. Anne's Court, Chertsey, UK

This house warrants selection for the sheer bravado in using circular planning in contrast with so much dreary grid-like plans that we are all familiar with. Although it would never be a prototype estate house, it has elements that are inspirational.

Stockbroker Gerald Schlesinger commissioned Sir Raymond McGrath,[26] who was a friend of Connell, to build a new house (1936–7) to replace an old one on the same site, known as Hill House at St. Ann's Court. Christopher Tunnard worked as consultant on the house and designed the landscape. Figure 1.21 shows the exterior of St. Ann's Court, a house of three floors with a plant room in the basement (Fig. 1.22).

The property has been described as a modernist 'Round House', which extends to approximately 6,500 ft^2, added to which is a converted nineteenth century 'Coach House' which provides a further area of approximately 4500 ft^2—situated within 20 m of the main house. These two buildings, both of which have been comprehensively and immaculately refurbished by the current owners, sit at the top

Fig. 1.21 Hill House, St. Ann's Court, Chertsey, Surrey, England. Sir Raymond McGrath. 1937. *Courtesy* The Modern House

[26]Raymond McGrath (1903–1977) was an Australian-born architect and interior designer who for the greater part of his career was Principal Architect for the Office of Public Works in Ireland.

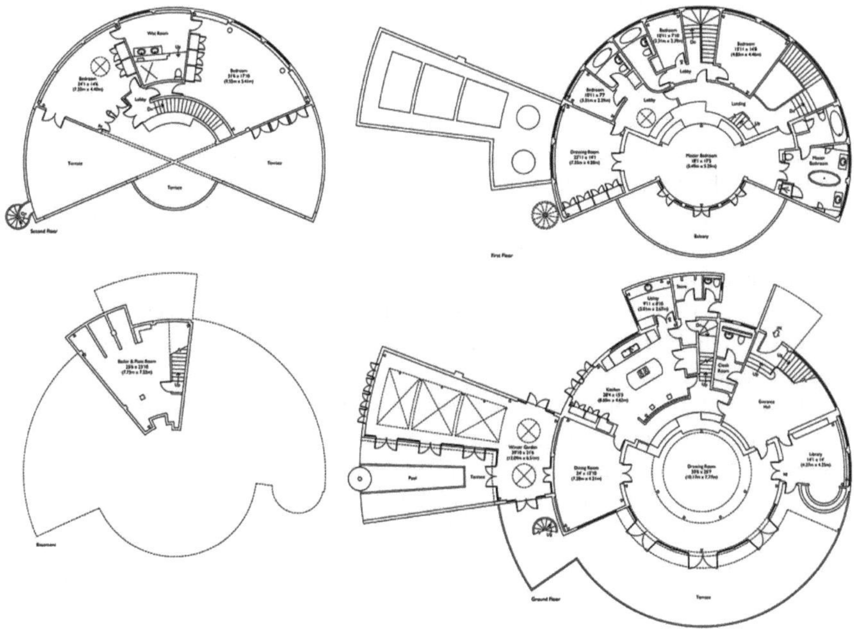

Fig. 1.22 Plans of Hill House

of an 8 acre site. This design is a brave departure from the nineteenth century's inherited past styles and joins the whole modern movement's will to explore new solutions. It has connections with Art Deco, which was fashionable during the decades of the 1920, 30 and 40s.

The geometric basis of the plan is concentric circles combining with a grid of diameters radiating out from the centre as a source for internal wall divisions. Upon closer examination, it appears that storage cupboards when fixed to these diameters do not actually taper to the centre but run parallel to the diameter wall. The circle enables the designer to provide a large sweeping curved glass wall with doors out on to the terrace, as well as the same for the master bedroom onto the balcony. There is also a semi-circular bay window to the library, a sort of spawned gesture from the mother circle.

I have shown some examples of housing, which are well beyond the social circle of public housing, but I think it is important to understand our heritage and the connections that are made. Both extremes will have the same functions of space to cater for sleeping, living, cooking, workspace, washing and utility. The major differences will be that the wealthier homes will have more space, a library, possible servants quarters, a his and hers sleeping and washing arrangements and numerous outbuildings with land to match.

1.5.3 Oliver Hill

This house is selected because demonstrates individuality in planning and encapsulates Art Deco brilliantly. One cannot ignore the prolific amount of work Hill did throughout the country as well as being linked to the glitz and glamour of the twenties and thirties and to the world of fashion, theatre and interior decoration. Hill began his career in the Arts & Crafts movement and worked as an apprentice to Edwin Lutyens.[27] In the 1930, his interests shifted towards modernism. This Grade II-listed three-/four-bedroom detached house in Fig. 1.23 is one such example of the Modern Movement on the Frinton Park Estate project, in which 200 acres of land was bought by the South Coast Property Investment Company, with a view to designing a new, modernist resort.

Whilst this house has some appeal with its white smooth curvilinear ice cream scoops upon closer examination of the plans, there appears to be a plethora of ill thought out bits. From Fig. 1.24, we can see the entrance is welcoming aided by the quadrant curved wall to the lounge, but then we enter a very tight hall and an even more tight WC. The lounge is inviting but steers us round to an ill-defined 'L' part, only to be confronted by a glass-curved external wall from nowhere squeezing us to

Fig. 1.23 Three–four-bedroomed house in Frinton-on-Sea. Oliver Hill 1934–5. *Courtesy* 'The Modern House'

---

[27]Sir Edwin Landseer Lutyens, OM KCIE RA FRIBA (1869–1944) was a British architect known for imaginatively adapting traditional architectural styles to the requirements of his era. He designed many English country houses, war memorials and public buildings.

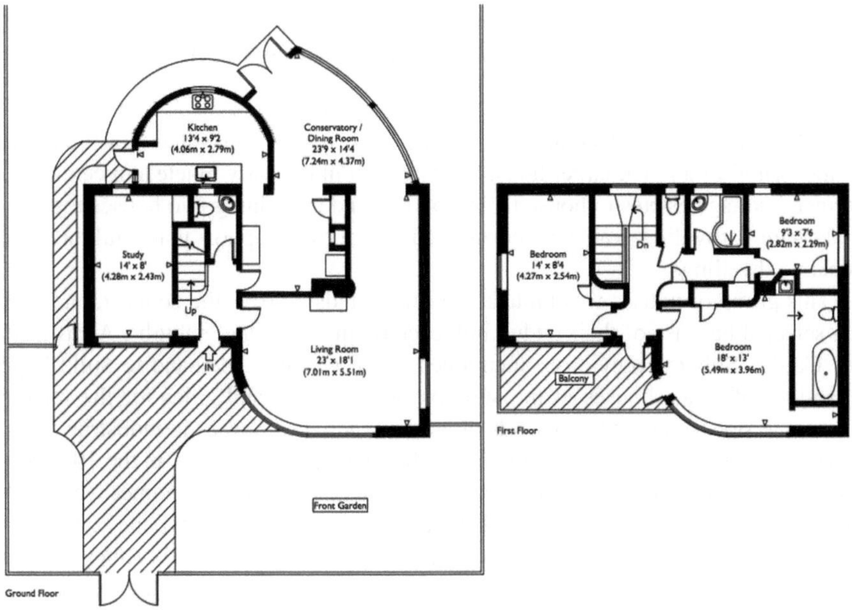

Fig. 1.24 Plans of Frinton-on-Sea house. A rhythmical plan

veer left in an uncomfortable manner. It is as though it is trying to echo the smaller curve in the lounge but badly. This is the dining room and a very asymmetrically unpleasant space it is. The semi-circular kitchen, which is tucked away out of sight, can never work properly as a kitchen no matter how pleasing the curved wall is. The central area is the utility room curiously open to view from the utility area. The first floor master en-suite has the washbasin stuck in the corner of bedroom, which seems to have minimal wardrobe provision. The flat roof over the entire house is certainly following 'modernist' principles by stubbornly resisting the traditional pitched roof which would be a little tricky here.

1.5.4 The Interwar Semi in the UK

So, immediately after the war, the 1919 (Addison) Housing Act introduced new standards for homes and together with the first Labour Government's 1924 Housing Act embedded the principle of subsidised council housing. Within about 15 years, under this legislation, a million modern, hygienic new homes had been built, many in suburban settings. Remarkable too was the growth in private home ownership. Those in work now had greater disposable incomes and the availability of affordable mortgages, from a rapidly increasing number of building societies, put home ownership within their grasp. **NHBC[28] Guide March 2015**

As I mentioned before (Sect. 1.2), the whole notion of a 'semi' is half of a house. It is a delusionary sales technique of making the occupier thinks that they are living in the 'wing' of a much larger house. They began in the early nineteenth century as farmers' cottage estates to house their workers, and combining two houses together that share a dividing wall was considered the cheaper option than building two separate dwellings.

This group of houses was built in their thousands all over the country, and they are selected because of their undue influence on image of the 'suburb'. Apart from the Garden Suburbs previously mentioned, there was huge demand to build large quantities of estate houses in the suburbs of our towns and cities for the middle classes, as opposed to the thousands of workers' terraced houses in industrial areas. These suburban areas became known as suburban 'sprawl' (coined in the 1950s) as they had no centre, focus or genius loci. Because land was cheap, homes were built at low densities with generous gardens and often in tree-lined avenues and open green spaces. Roads were not laid out as in America on a grid, but they were a mixed combination of curved avenues and straight roads sometimes producing annoying cul de sacs. The houses themselves would be a combination of two-storey detached and semis rather than the town terraced style. Their design often consisted of bay windows and arched brick porches with some odd Tudor half-timbered façade treatment echoing the grander country houses of bygone eras. Traces of Art Deco could be seen with the inevitable sunburst patterns on doors and gates as well as a lot of stained glass.

Due to the expanding motor industry, a new addition to the house was the 'garage'. Initially, this was seen as an addition like an extension because it had no prior precedent of being part of the architecture of the house. (See Sect. 3.2.4) And when they were built, they were seen as a status symbol, which became desirable even if the owner had no car. Eventually, they were designed as 'part' of the original house design and would sometimes incorporate a room above making it fit in line with the two-storey structure of the main house.

Figures 1.25 and 1.26 show a typical semi-detached of the 1930s with a total floor area of about 67 m^2. The entrance hall was large enough to have a side table, hat stand and sometimes a grandfather clock. Under the stairs would be panelled doors to a storage area. Fireplaces were still the main source of heating and consequently still a major focus of all rooms. The living and dining rooms would have been separate to begin with, but it soon became fashionable to knock through the dividing wall to create a more open-plan arrangement. The one and only bathroom, which bravely included the toilet for the first time, was adjacent to the kitchen to help economise on the plumbing services, although this imposed an embarrassing

[28]NHBC is the standard setting body and leading home construction warranty and insurance provider for new and newly-converted homes in the UK.

Fig. 1.25 Typical interwar semi-detached estate houses. *Photograph* Author

Fig. 1.26 Ground floor plan on left and first floor plan on right of a typical semi

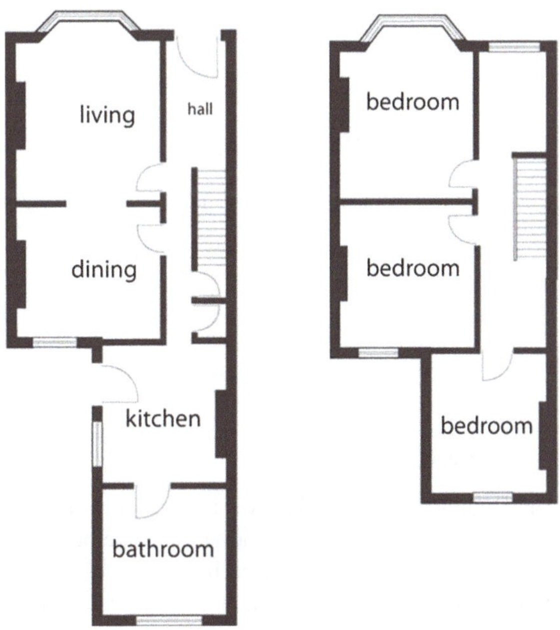

long walk to the bedrooms. The most common semi-detached houses were built around the mid- to late nineteenth century, although some argue that some were built in the eighteenth century and were often giving the appearance of one whole house as a means of retaining the grand status of a villa or a single dwelling as Fig. 1.27 shows. Streets were made of rows of these upper class 'estate' houses.

Fig. 1.27 Two semi-detached Georgian houses appearing as one dwelling. *Photograph* Author

1.5.5 Post 1940s Housing

The building industry was seriously affected by a shortage of skilled labour and materials, industrial capacity and manpower having been diverted into the war effort. The result was an acute housing shortage and, in order to alleviate it, a number of new methods of construction were developed.

Following the Second World War there was a surplus of steel and aluminium production, and an industry, until then geared up for the war effort, in need of diversification. These factors drove the move towards prefabrication. As a result, many new varieties of concrete (in both pre-cast and in situ forms), timber framed and steel framed systems emerged.

Non-Traditional Housing in the UK by Keith Ross of the Building Research Establishment (BRE). 2002

The Dudley Report of 1944 was an excellent update of the work of the 1919 Tudor Walters Committee and laid down improvements to improve the quality of life of residents. The Prefab—selected because it is part of an emergency post-war planning solution and the first example of prefabricated minimalist 'no frills' housing that does not fit into any evolutionary architectural style. It was born out of function, need and low cost. The whole country was in the grip of re-building and restoring people's lives that had been damaged by years of strife, human loss and insecurity. Because of the urgency of this re-build, the houses that were built were a combination of state funded (council housing) for rent and less ostentatious styles for the

Fig. 1.28 Typical post-war Prefab, The Excalibur Estate, Catford, London. *Courtesy* Municipal Dreams

private sector. There was also a demand for speed in the supply, and hence, the 'Prefab' came to be born. (Figure 1.28) Thousands of these prefabricated single-storey houses were built to minimum standards. The government stipulated that they should come under 'Temporary Buildings' and have a life of 10 years. All approved prefab units had to have a minimum floor space size of 635 feet^2 (59.0 m^2) and be a maximum of 7.5 feet (2.3 m) wide to allow for transportation by road. The Ministry of Works then held a public exhibition of five types of prefab at the Tate Gallery in London, in 1944:

- Two timber-framed designs, the Tarran and the Uni-Seco
- One steel-framed with asbestos panels, the Arcon.

One aluminium prefab, made from surplus aircraft materials, the AIROH (Aircraft Industries Research Organization on Housing) Another example was the 3031.

These houses retained a coal-fire, but it also contained a back boiler to create both central heating as well as a constant supply of hot water. The kitchen had a built-in oven, refrigerator and Baxi water heater. The kitchen and bathroom were built back to back to provide efficient distribution of services.

The plan below of a two-bedroomed bungalow shows how basic yet economic the planning was. The core service rooms of kitchen and bathroom were back to back. The back door was in effect a second front door for delivery of shopping goods (Fig. 1.29).

Post-war suburban housing continued to be built with rather more urgency but in a plain format than before especially if they were council housing as shown in Fig. 1.30. This resulted in a brick plain box punctuated by windows and doors and a tiled roof with chimney.

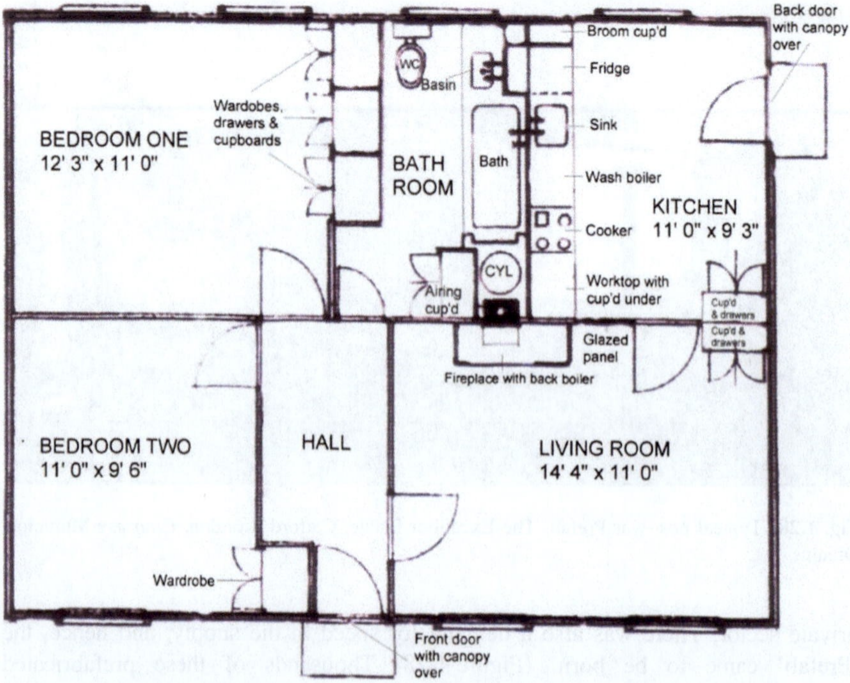

Fig. 1.29 A typical Prefab plan

Fig. 1.30 Suburban Council estate housing built after the war. *Photograph* Author

1.5.6 Harlow New Town, UK—Planner Sir Frederick Gibberd

This house was selected because it represented a departure from the suburb to a 'new town' with its own heart and neighbourhood facilities. The 1946 New Towns Act was established to cater for post-war demand but rather than expand existing towns with suburban sprawl, these new towns would be built beyond the green belt[29] and have their own centres, shops, workplaces and leisure pursuits. Examples were Stevenage, Hatfield, Bracknell, Hemel Hempstead, Basildon and Harlow amongst others. Eventually, people would move out of the bigger cities to these new towns, which provided cleaner air, less traffic and a healthier lifestyle.

Unfortunately, whilst there was more green space surrounding these houses, this did not mean more internal space inside the house. The total floor area of the Harlow house in Figs. 1.31 and 1.32 by Cadbury Brown[30] is approx. 72 m^2 which

Fig. 1.31 Harlow housing, *Courtesy* Municipal Dreams

[29]Green Belt: In UK town planning, the green belt is a policy for controlling urban growth. The idea is for a ring of countryside where urbanisation will be resisted for the foreseeable future, maintaining an area where agriculture, forestry and outdoor leisure can be expected to prevail.

[30]Henry Thomas Cadbury-Brown RA (1913–2009) was a British architect. He was educated at the Architecture Association where he was influenced by the architecture of Le Corbusier and Walter Gropius. After graduating, he worked for architect Ernő Goldfinger and became his lifelong friend. He went on to set up his own successful practice.

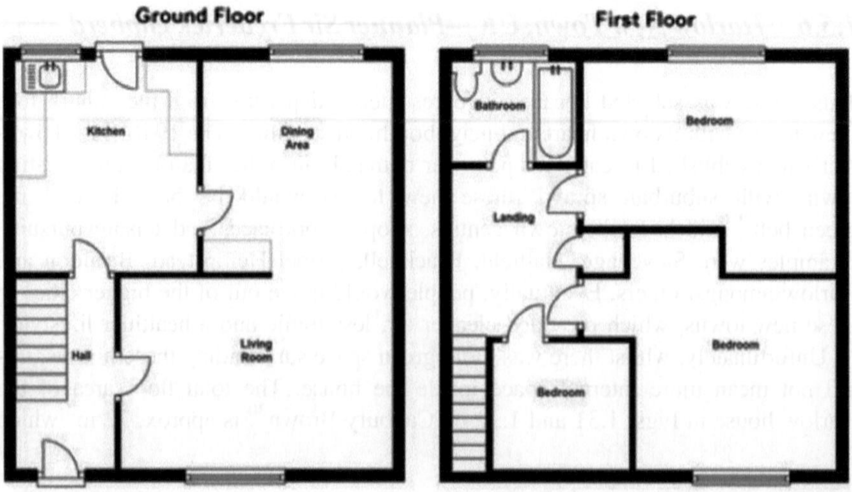

Fig. 1.32 Plans of Harlow house

just satisfies the future Parker Morris[31] standards of 1967 (abolished in 1980 by Margaret Thatcher), but as you can see from the floor plan, it is pretty dire apart from being a boxey soulless place to live. I think that the predominant thought with the architects was how the elevation looked as a modernist statement rather than considering more the needs of the occupants. This was the beginning of loosing the independent house to a collective terraced structure.

The living room is open to the dining room creating a more interesting 'L'-shaped arrangement. The kitchen still provided the only entrance to the garden as was the case in Victorian times. There is still only one bathroom including the toilet. The 'third' bedroom is particularly mean and could sometimes be referred to as a 'box' room.

1.5.7 Suburban 1950s Detached House—191 m^2. Milton Keynes. UK

This house is selected because it echoes the 'Moderne' style but is planned with more space provision and hence falls within the luxury band. My own design for a five-bed house in Figs 5.56 and 5.57 is 146 m^2.

This more expensive property is in Church Green Road in Bletchley now part of Milton Keynes. The name Church Green Road refers to the nearby twelfth-century

[31]Parker Morris Standards of 1967 dealt with space, access and ergonomic requirements and were to be applied to all housing as mandatory whether publicly or privately funded. These ceased to be mandatory in 1981.

Fig. 1.33 Five-bedroomed detached house built 1950. Milton Keynes. *Courtesy* Michael Graham Estate Agents

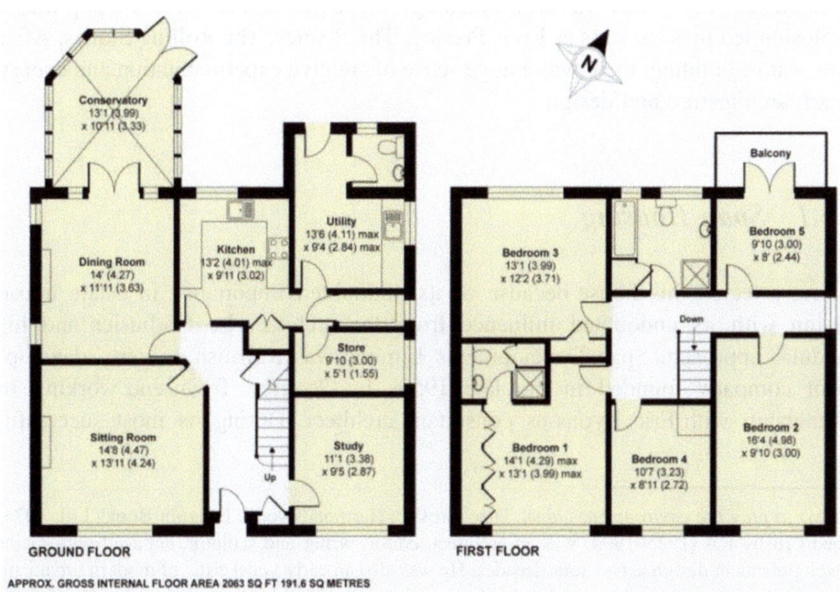

Fig. 1.34 Plans of 1950s detached house. *Courtsey* Michael Graham Estate Agents

parish church of Saint Mary and its Green, which forms part of the West Bletchley conservation area, which includes the now famous Bletchley Park, home of the WWII code breakers. Well-kept Victorian terraces, a converted Victorian school and schoolhouse, post-war properties of all sizes, and a number of listed thatched cottages make up the tree-lined street scene. Opposite two of these thatched properties stand this detached property, built in 1950 by well-respected local builders 'Tranfield'. The plot is almost 100 ft. in length with landscaped front and rear gardens. Built in the 'Moderne' style, this is a spacious but rather old-fashioned house. Good to see a good-sized utility room plus WC as well as a store room on the ground floor. On the first floor, the en-suite bathroom, probably a later addition, is uncomfortably small at the expense of taking space away from the master bedroom (area 17 m^2). Two of the bedrooms are extremely undersized and could be described as box rooms.

1.6 Modern Housing from 1960

An 'interesting' environment is characterised by a high level of aesthetic information, using a large vocabulary of different elements. Modern buildings, lacking associative potentiality, offer particularly low quantities of information and lose our interest in the first simplification.

Theo Crosby.[32] **'How to play the environment game'.**

By the mid-1950s, a third of all households had television. Interior styles were abundant with production of textiles, furniture, lighting and wall-decorative materials. Kitchens and bathrooms were produced revealing progressive design in their components. The 1960s was an awakening of a sexual revolution and a pop music explosion led by such stars as Elvis Presley, The Beatles, The Rolling Stones. After post-war re-building, there was a huge sense of creative experimentation and energy in art, architecture and design.

1.6.1 Span Housing

I have selected this house because of its undoubted importance in estate house design with its undoubted influence from the architect Le Corbusier and his modular approach. Span Developments Limited was a British property development company founded in the late 1950s by Geoffrey Townsend working in partnership with Eric Lyons as consultant architect. During its most successful

[32]*How to play the environment game.* Theo Crosby. Harmondsworth. Penguin Books Ltd. 1973. Theo Crosby RA (1925–1994) was an architect, editor, writer and sculptor, engaged with major developments in design across four decades. He was also an early vocal critic of modern urbanism. He is best remembered as a founding partner of the international design partnership Pentagram, and as architect for the reconstruction of Shakespeare's Globe in London.

Fig. 1.35 Three-storey Span four-bed terraced housing in Weybridge designed by Eric Lyons in early 1960s. *Courtesy* The Modern House

period in the sixties, Span built over 2000 homes in London, Surrey, Kent and East Sussex—mainly two and three bedroom well designed single-family homes and apartment buildings. The emphasis was provision of facilities in the most compact form. I had the pleasure of meeting Eric Lyons when I was a student and he told me that he had great difficulty persuading high rise (tower blocks of flats) building developers to get rid of the fireplace which was now redundant with modern central heating systems (Fig. 1.35).

The surprising projecting feature by the entrance is the ground floor WC as seen in Fig. 1.36! The hallway is small and so is the one reception room with no dining area provided. The kitchen is very tight especially if two people are in there. With only one bathroom on the second floor serving four bedrooms, my conclusion is that this design is impoverished. Yet its total floor area is about 96 m^2 but take one bedroom away to compare with other three bed houses; it would be about 84 m^2. Considering the serving facilities, the four bedrooms are far too many, and hence, Lyons has squished the basic footprint and added a third storey, which is unusual for a house that does not gain a great deal.

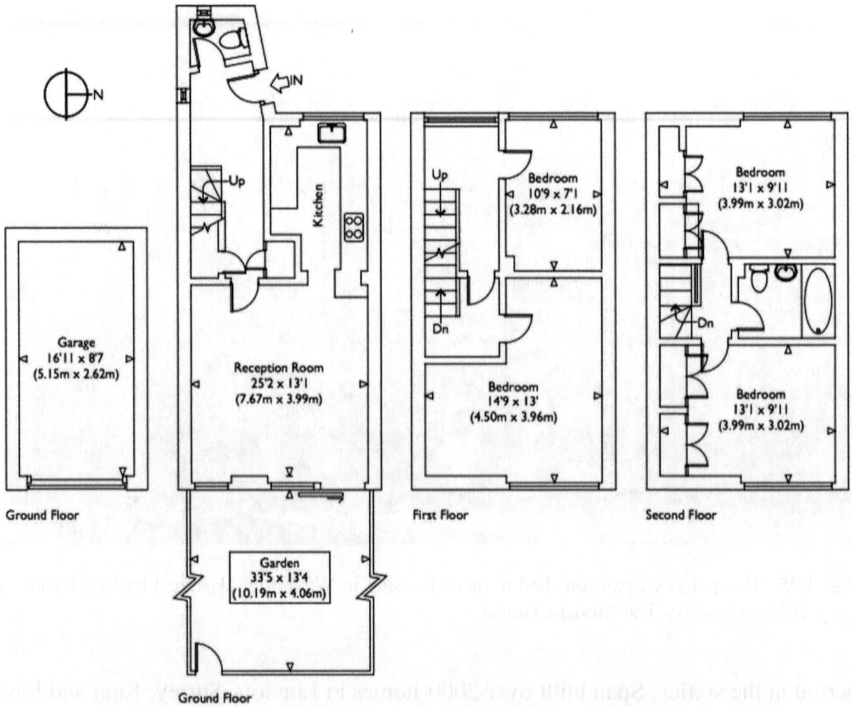

Fig. 1.36 Plan of terraced span dwelling. The garage is in a separate block

1.6.2 New Ash Green

The concept of an entirely new village built in the middle of the Kent countryside drew a lot of opposition from various bodies including the locals from the surrounding villages including my own relatives who lived in one of those villages. The plan was to not only house London overspill but provide more housing for people wishing to live in the countryside instead of imposing a strain on existing villages by rapid expansion. As it turned out, the demand has been so great that the other villages have had to cope with more houses being built anyway. The main opposition point of view was that villages in this country have grown organically over centuries and to suddenly build a new one without history was madness, brave but mad. At the same sort of time, a brave new town was being planned called Milton Keynes in Buckinghamshire. See page Sect. 2.2.1.

A plot of land that had once formed part of both North Ash Farm and New House Farm was bought in 1961. The name New Ash Green was derived from these two farms. The nineteen 'neighbourhoods' of New Ash Green were to be composed of 100–150 houses built in varying densities. Span were again responsible for building over 50% of the village using Eric Lyons as the lead architect. Being personally close to the building of this village, I can report that after the first few phases of building it was realised that the village had no pub! Villagers would have to go to the nearby villages for a drink which has since been rectified over the years, but it does reveal a huge misunderstanding into what makes the heart of a village: a pub, a church and a shop, and if by demand: a school. In the early 1970s, I helped design its first bank! As time progressed, the shopping centre fell into disrepair due to overseas landlords abrogating their duties. This is the kind of situation that is happening in London and elsewhere, and something has to be done about it.

The total floor area of this house (Fig. 1.37) is approx 130 m^2. The plan in Fig. 1.38 reveals a curious undulation of the envelope which is difficult to understand. The plan may have some cohesion as a shape, but the interiors would feel decidedly odd. Lyons continues with the small entrance hall as he did with the Span houses. Jammed in next to the wc is a shower which would test the manoeuvrability of anyone. The kitchen is only accessed via the hall, and the garden room must be the dining room. The living room is denied vision or access to the garden but has a doorway into the study, which does have access to the garden. The emphasis here is not so much on open plan and sharing, but rather privacy for all rooms. To have five bedrooms with no en-suite and one-shared bathroom and an old-fashioned separate toilet is inexplicable. Why does only one bedroom have the privilege of having built-in wardrobes?

Fig. 1.37 Five bed end of terrace house in Lombardes, New Ash Green designed by Eric Lyons 1968. *Courtesy* The Modern House

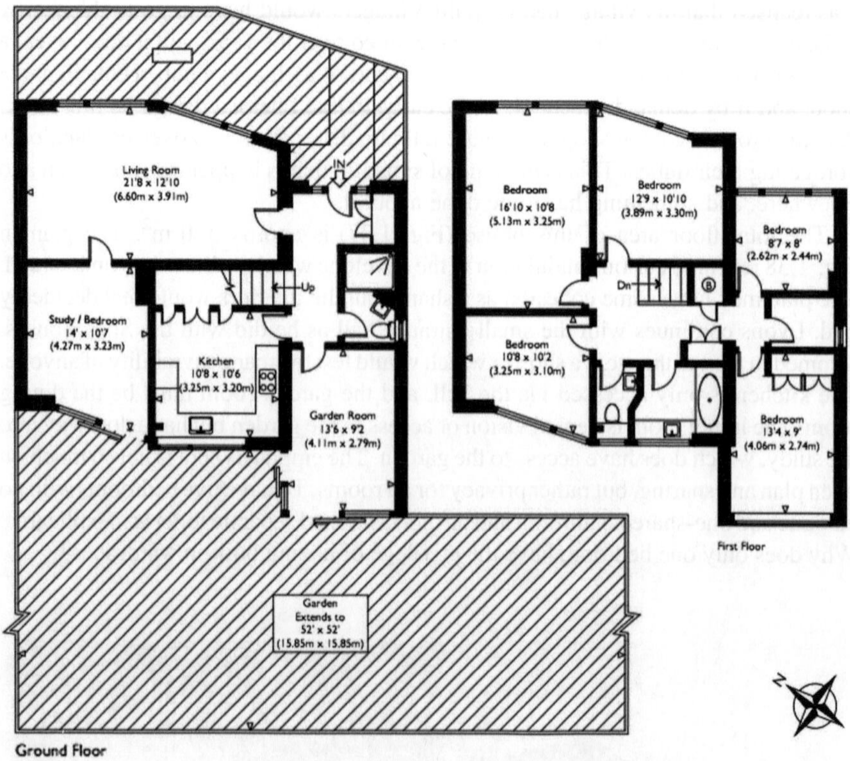

Fig. 1.38 Plans of New Ash Green house

1.6.3 World's End Flats, Chelsea, & RCA Building, London

I know this is not in the 'house' category, but the design suits my argument in Chap. 5. When the World's End housing estate was completed in Chelsea in 1977 after 10 years of construction, it was deemed a failure. Three decades on, the enduring excellence of the design by Eric Lyons and HT Cadbury-Brown is recognised by architects and the people who live there. The polygonal planning was unprecedented in Cadbury Brown's work although he had used it in Essex University lecture halls and the common room and library block for the Royal College of Art. Having been a student at the RCA in the 1960s, I especially enjoyed working in those spaces. The shapes had a natural feel to them as you can see from the plan in Fig. 1.39. Little did I know that the polygonal form would play an important part in the later section of this book. This building uses the hexagon as the main geometric basis for planning.

The plan in Fig. 1.40 shows two-bedroomed World's End flats both identical apart from the balconies position: the upper flat's balcony is off the dining/kitchen space, and the lower flat's balcony is off the living room. Apart from the rectilinear

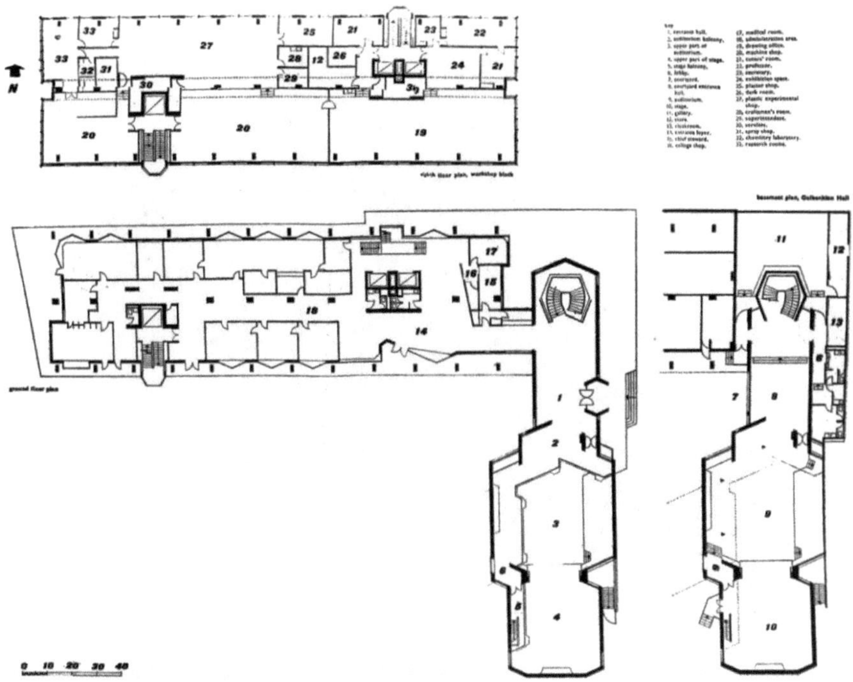

Fig. 1.39 Plans of the Royal College of Art, London. Designed by H. T. Cadbury-Brown and Hugh Casson 1962. *Courtesy* Architectural Review October 1962

Fig. 1.40 Plan of World's End flats designed by Eric Lyons and H T Cadbury Brown. *Courtesy* BDonline. These two-bedroomed flats are separated horizontally down the centre

Fig. 1.41 Exterior of World's End Flats. *Photograph* Author

central servicing core of bathrooms, toilets and storage, every other room has at least one wall going off at 45°. Portions of an octagon are used. This reduces the 'box' syndrome and is a huge emotional release in those spaces. It is interesting to note that contrary to current common practice, the kitchen/dining area is separated from the living space, which actually has enough room to have a second table for dining or other activities. The undulating plan is expressed on the exterior to provide a more stimulating tower form than many of the tower blocks that have been built in our towns and cities (Fig. 1.41).

1.6.4 Windows

If we take a quick survey of window design since Tudor times and more particularly how the windows are framed, we can see how their existence was initially expressed as being very important to the architecture of the building. The function of letting daylight and fresh air in, and being able to see out, places huge importance upon the window. As Fig. 1.42 shows, by the present day, the framing has been reduced to almost nothing. When you have glass walls, this is eliminating the window altogether and it does not say much for a solid wall either! (See Chap. 6 p. 113 of my book 'Interior Design: Theory and Process').

Hence, in Chap. 5 my proposed house design expresses the windows and doors with a strongly assertive frame both externally and internally.

Fig. 1.42 A summary of window framing through history. *Drawn* Author

1.7 Conclusion

I hope the reader can understand why I am focusing on the planning of these dwellings irrespective of the social conditions and political and economic forces that were taking place in each era. I was a student in the 1960s, and I can remember how many of the type of houses I show in this chapter were praised for their imaginative break from the traditional semi as shown in Figs. 1.25 and 1.30. It is a pity that the interior layouts did not match the rapture the architecture received which was under the domination of modernism. As my next chapter reveals, the interiors of modern houses built today also suffer from an architectural domination at the expense of interiors that work for the occupants. The following is an extract from Building Design dated 2.12.15 beginning with a quote from the RIBA:

> *In 1920, the average semi-detached new-build was four bedrooms and 1647 sq ft, (148sqm) according to a survey by the* ***Royal Institution of British Architects (RIBA)****.*
>
> *The equivalent now is just three bedrooms and 925 sq ft (86sqm). Typical new terraced houses have, meanwhile, shrunk from three bedrooms and 1020 sq ft to two bedrooms and 645 sq ft (58sqm).*

The 1920 4 bed house area is 148 m^2 which is much more than my final proposal later in the book in Chap. 5 of 126 m^2 as a minimum for a three bed house. Of course, I am competing with the national standard recommendation of 93 m^2, which I maintain to be ludicrous. Why have houses become smaller? The answer to this question will be unravelled in this book but put simplistically, since 1920 we have more people to accommodate and available land space has become scarcer and more expensive. We could not build the 1920 house now with the same materials and techniques as it would be too expensive, and it is likely it would not pass current building regulations. A radiator was a rare sight in 1970 when just one-quarter of homes had central heating. By 1990, this had risen to nearly three quarters. So modern technology has helped builders and developers to build at a reasonable cost to meet today's demands of eliminating condensation and damp, improving insulation and reducing fuel bills, but of course, as this book repeatedly confirms, they are failing to provide a proper product.

The Building Design extract continues with the theme of not enough space provision:

> *'More than half of the new homes being built today are not big enough to meet the needs of the people who buy them, the RIBA says. A new study of space standards reveals that a squeeze on space is depriving thousands of families of the room needed for them to live comfortably. Called Homewise, the survey says that on average buyers of a new three bedroom home are missing 4 sq m – the size of a family bathroom. It adds that the smallest three bedroom homes surveyed by RIBA are missing space equivalent to an entire double bedroom. Homes in Yorkshire are by far the smallest in England – the average new three bedroom home in Yorkshire is 25 sq m smaller than one in London. At only 84 sq m, the average new home in Yorkshire is smaller than one on London by the equivalent of a double bedroom and a family living room. RIBA is calling for national space standards on all homes and its president Jane Duncan added: "We urgently need new homes, but building small homes or cutting corners when converting office buildings to flats is short-sighted and fails the people these new homes are meant to serve. The Government must take action to ensure a fairer minimum space standard is applied to all new homes across the country.'* **Building Design 2.12.15**

Hence, my plea in this book for the space standards is to be reviewed and re-assessed to reflect modern living and the needs of families. Despite the Housing Acts of 1964 and 1985, and various reports by Karn and Sheridan, and the Housing Corporation's (HC) 'Scheme Development Standards' (SDS) in 1993, which seemed to address such issues as overcrowding in what seemed to be a free for all undisciplined occupancy of houses, these did not prepare authorities for the present task that now confronts us. This implies that more research is required directed to this end as the President of the RIBA Hodder recommends:

> *We need further research into what makes good housing and indeed what makes good architecture in general, to facilitate the wellbeing of those who use our buildings and to advance the impact of our profession* Stephen Hodder, **RIBA President**.[33]

---

[33] Stephen Hodder, RIBA President. RIBA Home Improvements: Housing Research and Practice, October 2013.

Chapter 2
Case Studies

Abstract My selection consists of house types of varying categories, interiors that demonstrate imaginative solutions to adjustable and moving parts, and creative furniture solutions. So the survey spans from architectural considerations through to interiors as one cannot be appraised without the other. Whilst my focus is on mass estate housing as opposed to 'social or affordable housing' I have found many examples of house design that are in other categories but are nonetheless inspiring, well designed and can teach us something. Hence, my survey is deliberately broad simply because there are not enough estate-type houses worthy of inclusion. Out of 22 houses listed here, 12 are estate type and 6 are with integrated garages. The traditional fireplace with projecting chimney breast has almost totally disappeared to be replaced by central heating and the odd wood burning stove. I should point out that some of the schemes shown here could also consist of mixed development of apartments and some social housing. Most house schemes will have a small key plan diagram summarising one of the six types of plan with rooms indicated in orange. Stairs may vary. The type of house construction is stated after each case study heading. The technical information is not standardised because it is in the form that the architects and designers have given to me. I show some examples of apartment interiors and some furniture systems because their designs have some merit in terms of imagination and creativity, something that is lacking in the estate house type. There is the question of to what degree is an estate house equipped with fitted furniture items? The standard supply is reduced to mainly the kitchen and bathroom areas with built-in wardrobes in some bedrooms. The rest of the house consists of characterless empty boxes that are left to the occupier to equip and furnish. This exposes to me a complete lack of understanding by the designer/builders of what the basic shell should consist of.

A. Sully, *The Estate House Re-designed*,
https://doi.org/10.1007/978-3-319-90397-2_2

Fig. 2.1 Mock Tudor and a common characterless example in UK

In terms of house style, it will be seen that, similar to examples in Chap. 1, there are designs that break away completely from any signs of tradition to those that still embrace certain traditional principles (Fig. 2.1).

The above house types are perhaps the worst examples of estate housing today.

2.1 Areas

2.1.1 Recommendations

One means of comparison of these projects is to focus on the master bedroom area provided. I have produced my ideal minimal plan area 14.8 m^2 for a master bedroom (Fig. 2.2), and I intend to test this against most of the case studies listed. Note my inverse corners as this suggests that a room can demand a shape rather than fit into a rectangular grid. Otherwise, the 90° corner remains an unusable space. It is obviously symmetrical giving both partners an equal amount of space and storage. Plans like this should influence the architecture of the building in deciding where windows should be instead of a building being planned without any detailed thought of the internal layout of each space. UK Building Regulations 2015 states that the minimum area should be 11.5 m^2 which I obviously disagree with even though the Tudor Walters Committee (see Sect. 1.4.3) recommends 14 m^2 which is close to my own. Table 2.1 is the recommended space standards table for the whole house taken from the current UK Building Regulations. The average new home in the UK is 76 m^2; in Ireland, it is 87.7 m^2, and in Denmark, it is 137 m^2.

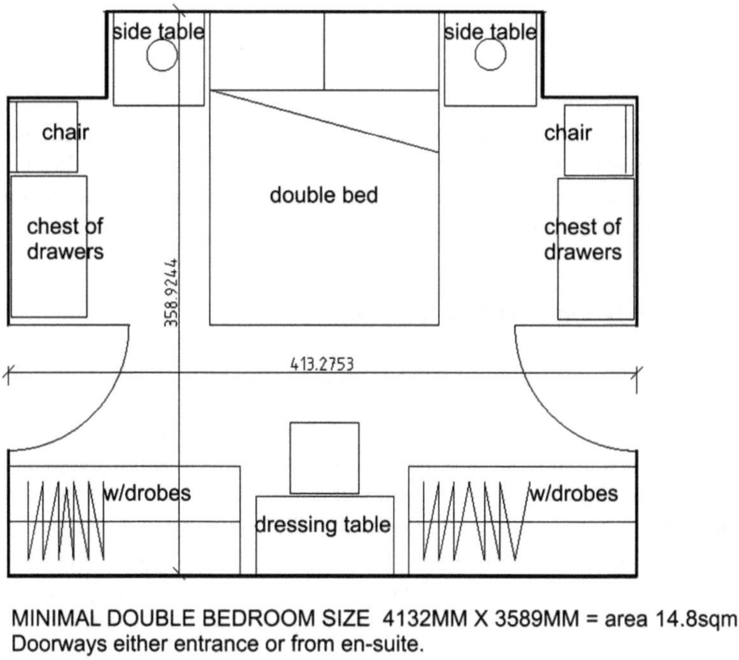

Fig. 2.2 My ideal minimal master bedroom plan of 14.8 m^2

Table 2.1 Minimum gross internal floor areas and storage (m^2). Space standards

Number of bedrooms (b)	Number of bed spaces (persons)	One-storey dwellings	Two-storey dwellings	Three-storey dwellings	Built-in storage
1b	1P	39			1.0
	2p	50	58		1.5
2b	3p	61	70		2.0
	4p	70	79		
3b	4p	74	84	90	2.5
	5p	86	93	99	
	6p	95	102	108	
4b	5p	90	97	103	3.0
	6p	99	106	112	
	7p	108	115	121	
	8p	117	124	130	
5b	6p	103	110	116	3.5
	7p	112	119	125	
	8p	121	128	134	
6b	7p	116	123	129	4.0
	8p	125	132	138	

2.2 Mixed Development

As more money became available for building private houses following the removal of building licences in November 1954, features that became popular included staircases rising from the living room and open-plan living areas. Prior to this change, staircases had to be enclosed in hallways and landings separate from other rooms for means of escape in case of fire.

2.2.1 Oxley Woods, Milton Keynes, UK

Eco Homes Development, 2008. Architect: Rogers Stirk Harbour. Prefab steel frame.

Oxley Woods is a development of 145 homes of innovative and sustainable design. It consists of a number of different house types to suit single people, couples and families. The overall site, which is in Milton Keynes[1] covers 3.26 ha. 25 of the 145 houses have received funding under the first time buyers initiative run by English Partnerships to assist people into home ownership. Unfortunately, in the last few years, there have been technical problems with the houses leaking, and there are disputes going on to resolve matters. Floor area for a three-bedroomed house is 81 m^2 (Fig. 2.3)

Fig. 2.3 Oxley Woods, Milton Keynes, 2008. Architect Richard Rogers. *Courtesy* John Markus

[1]Milton Keynes. A new town in Buckinghamshire designated in 1967 to cater with London population overspill.

The developer, George Wimpey, has worked with design partner Rogers Stirk Harbour & Partners (RSHP) (formally Richard Rogers[2] *Partnership) to produce a development that meets high environmental standards, within strict budgetary controls. Off site manufacturing has been used extensively. Prefabrication in a factory has been used for the main components of walls, ceilings and floors. In addition, other components (such as stairs) and service installations (heating, ventilation and power) have, despite being complex to achieve the required environmental standards, been largely standardised and again prefabricated off site. These latter elements, linked to bathrooms and utility spaces, form a core 'service zone' within each dwelling, allowing a flexible and adaptable arrangement for the 'living zones' (living, dining and bedrooms) around the core.* Lifetime Homes

These houses appear to be visually striking, inventive as well as being designed by one of our leading architects Richard Rogers. I am absolutely staggered to find an internal plan that is beneath human dignity. The small hall is open into a space that is the kitchen, the dining area, and the living area. No way Hosé! How can this be when the development has been hailed (when it was first built) with praise for achieving Eco Home status! To force the toilet to share utility functions exposes potential conflict of use. Two of the larger bedrooms are undersised with no built-in wardrobes and hence unworkable. Area of master bedroom is 12.5 m^2 (Fig. 2.4).

TECHNICAL

The 145 properties have been flat packed constructed from sustainable materials, using insulated steel panels overlaid with cedar cladding, and employing

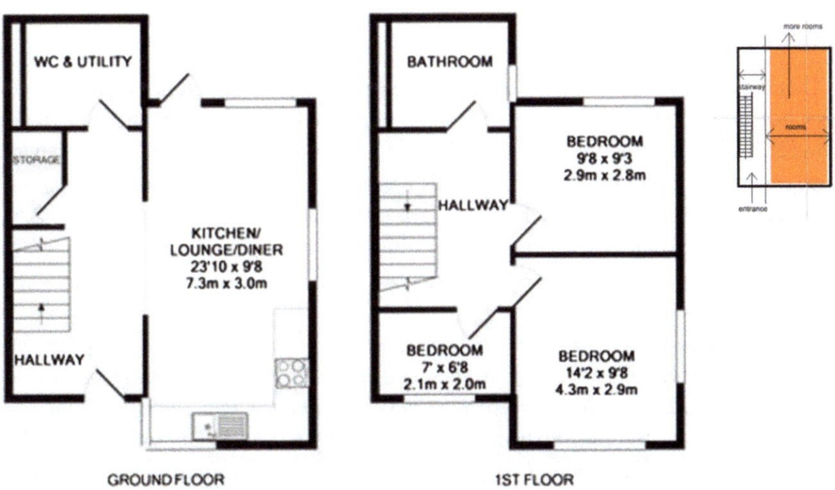

Fig. 2.4 Plans of an Oxley Woods three bed house. Re-drawn by author

[2]Richard George Rogers, Baron Rogers of Riverside CH FRIBA FREng RA (born 1933) is a British architect noted for his modernist and functionalist designs in high-tech architecture. He is best known for his work on the Pompidou Centre in Paris and the Lloyd's building and Millennium Dome both in London.

unique features including an 'EcoHat'—allowing hot air to be reused to optimise energy consumption and provide passive solar water heating—as well as exceptional layouts to maximise space and light. The houses were designed to exceed 2002 Part L Building Regulations, meeting Code Level 3 in the Code for Sustainable Homes. It is predicted that the good insulation and airtight design of the homes, in conjunction with the low energy ventilation system of the EcoHat, will significantly reduce the energy required to heat and cool the homes compared to standard homes. In the summer, the ventilation system can be used for night-time cooling. The low-watt fan actively ventilates the house and filters out pollen. The EcoHat is 'future-proofed' allowing for further adaptations; existing components can be exchanged for new ones. For example, an optional heat recovery mechanical ventilation system can be put into take the dwelling up to level 3 of the Code for Sustainable Homes rating and adding triple glazing to the properties would take them up to level 4, whilst local energy generation would enable the development to meet level 5.

2.3 Terraced Houses

2.3.1 House—New Islington Development, Manchester, UK—2016

Timber frame

This remarkable new award-winning development is the best that I have seen for mass housing. The architects are Shedkm who have done some beautiful work elsewhere, and they had an enlightened client Urban Splash. The key to this design is customer participation whereby choices are on offer with regard to planning and finishes. It is a shame that they have straight staircases instead of dogleg (Fig. 2.5).

This is such an intelligent way of operating by working alongside purchasers to ensure they are satisfied with the final outcome. They will also feel a stronger sense of ownership.

The designers write: *hoUSe in Irwell Riverside, a new neighbourhood on the edge of Manchester City Centre. Riverside two & three storey family homes set within acres of landscaping. House is a new concept for mass housing that is a direct challenge to the established typologies of the major UK housebuilders. We have taken the familiar pitched roof form and turned it longitudinally to create a long 'loft-style' spacious home that is aimed at the urban house-buying market. The raw shell of the house contains simply a stair, a kitchen and bathroom pod. A series of layouts can be selected with a choice of number of bedrooms together with a choice of finishes. The house is factory-built to order.*

The ground floor in Fig. 2.6 offers a one wall set of kitchen units in plan 'A', increasing kitchen size with additional island unit in plan 'B', an 'L' shape kitchen in plan 'C' which also creates a hall and possibly a study/play room.

Fig. 2.5 House two-storey two bed house by Shedkm in Manchester, 2016. *Courtesy* Shedkm

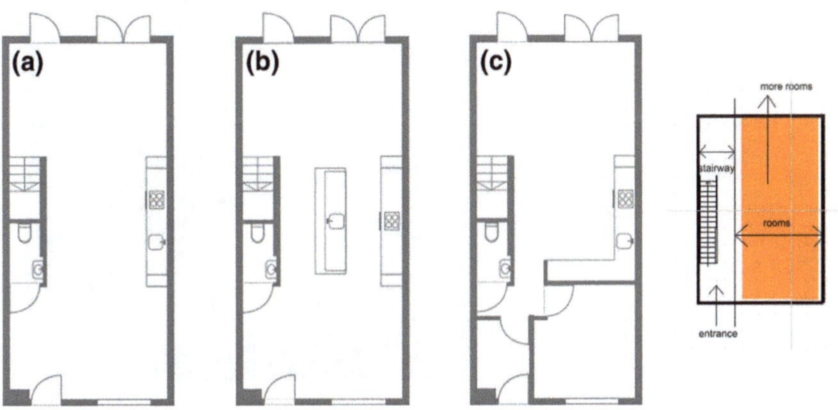

Fig. 2.6 House—Optional layouts for the ground floor kitchen

The first-floor options in Fig. 2.7 are quite simple with either two large bedrooms or one large, one small and a box room or study. The bathroom stays the same.

Figure 2.8 shows the simplicity of form awaiting the furnishings of the occupant. The pitched roof creates height which relieves oppression, and the angle is most welcome (see my Introduction). In older houses such a pitch would only be exposed by removing a ceiling and joists to what may have been a loft. I have often said that modern housing estates have stripped away any form of interior design to such an extent that the only sign of humanity left is the skirting board! And yet here, we

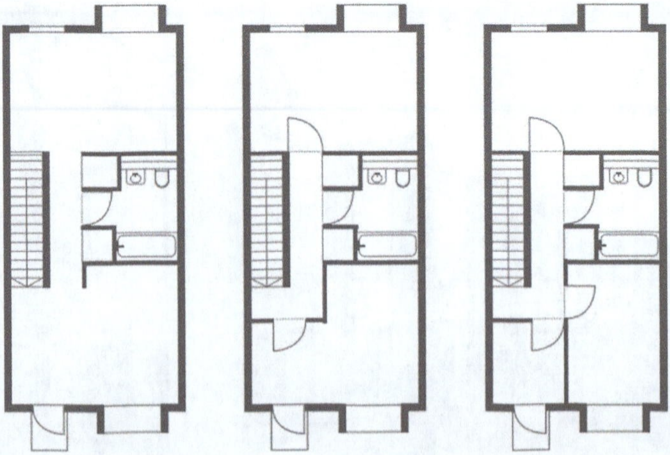

Fig. 2.7 Optional layouts for first floor bedrooms

Fig. 2.8 View of kitchen area showing 'loft' style roof

have a project that has got rid of the skirting board! But the spirit of the design and its architecture is enough of a strength to provide human comfort.

Floor area—Two storey—93 m^2 or three storey—139 m^2.

TECHNICAL

External walls

SIG insulated timber frame overclad with Equitone Tectiva panels (estimated U-Value 0.1)

Roof
SIG insulated timber frame finished with single ply membrane (estimated U-Value 0.1)

Windows and external doors
Windows—Ideal Combi Futura
External Doors—Ideal Combi Futura

Floor finishes
Customer choice of floor finishes, engineered oak or carpet (subject to additional cost)
Stair finish will match the floor finish adjacent to bottom step:
Solid oak finish to stair where engineered oak chosen
Carpet finish to stair where carpet chosen
Softwood stair where no floor finishes chosen.

Wall finishes
Painted MDF panelling to internal partitions
Painted dry lining to ceiling

Heating and hot water
Slimline, ceramic, electric panel heating
210 l storage tank

Lighting
LED/low energy lighting throughout

Media
CAT 5 wiring throughout
Broadband connection facilities and satellite points

2.3.2 Barratt Homes Development, Houndwood, Somerset, UK 2016

Brick and block

The supply of mass public housing is usually in the hands of a few developers, and Barratt Homes is one company that leads the way. Their newest development is in Houndwood in Somerset, and I am showing details of a four-bedroomed terraced house listed for sale. I am afraid that, despite seductive exteriors with varied textural modelling, it attracts the sort of criticism that people have stated in my Introduction (Figs. 2.9 and 2.10).

If the garage is removed, you will see how small the overall area of the ground floor is. An integral garage is a slick marketing deception of making the house appear larger than it is. The Building Regulations recommends an average area for a three bed to be 93 m^2, whereas this Barratt home is under 60 m^2 (excluding the integral garage). The general lack of storage is a social crime. In the past, staircases were not allowed to open onto a living space but had to pass through a hallway to

Fig. 2.9 'Nashira' a new Barratt four bed terraced house, Houndwood Somerset. 2016. *Courtesy* Barratt Homes

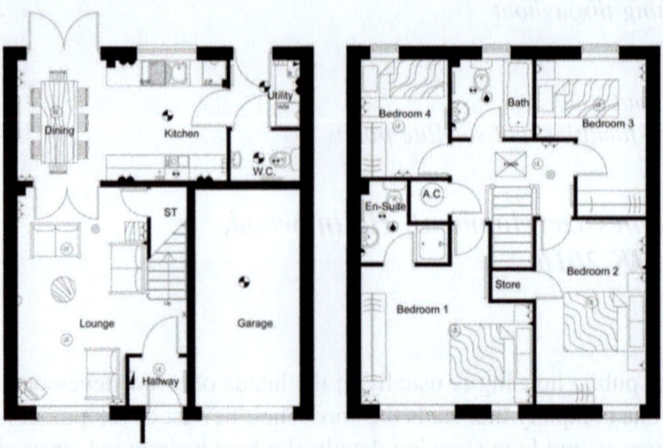

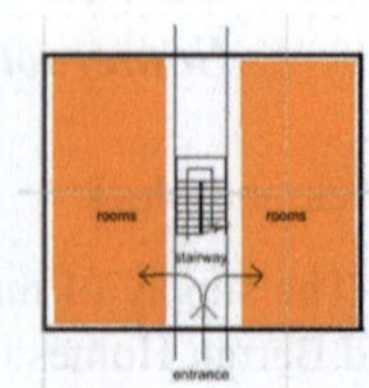

Fig. 2.10 Floor plans of 'Nashira' a new Barratt four bed terraced house, Houndwood Somerset. 2016. *Courtesy* Barratt Homes

satisfy means of escape in case of fire. The regulations now allow these to open into a living space, and consequently, they impinge upon it as does the ludicrously small hallway. The noise made in the lounge can go straight up the stairs to the bedrooms as there is not intervening door. The hallway has no decent room for the amount of coats and shoes that five people would have and is planned as an ugly intrusion into the lounge. The lounge has minimal wall space for book shelving and is extremely tight for people movement as well as feeling generally cramped for space. Because the kitchen is so small, the dining table would be used for food preparation, homework and general writing laptop work. Once people are sitting at the table, it will be a squeeze to go through from the lounge to the kitchen. As waste demands segregation of rubbish into at least three containers, I see no space or provision for this. The utility room is hardly a room as it just has space for a washing machine and a drier with the wall-mounted boiler above. No space for ironing and how do you manage the laundry?

As I have mentioned in my previous books, a staircase should be a dog-leg for safety reasons as to minimise the falling distance. Sixties or so treads as a single straight staircase is a constant source of accidents, and I am staggered that with all the current emphasis of health and safety that this has not been looked at (see Sect. 3.1.2).

If we look at the first floor, it is so cramped as to be unbelievable. Off the landing, there are four bedrooms, one bathroom, an airing cupboard door, a loft door and the staircase entrance itself. It is so claustrophobic and difficult for two people to pass each other. The airing cupboard is way too small and will not store the required linen of a house this size. The width of Bedroom 4 is the length of a single bed, which makes it pretty tight, and there is no room for a chair to put clothes on, nor space for a chest of drawers. This would in all probability be a child's bedroom, which would be unworkable as there is no room for a desk. Bedroom 3 is a larger single bedroom but still tight to accommodate a chair, chest of drawers and a desk. The master bedroom will be a squish if two chairs and two chests of drawers have to be accommodated. The en suite is a joke but enough is enough as I will not be buying it! Area of master bedroom is11.7 m^2.

2.3.3 Villa Verde, Chile—484 Housing Units 2010–2013

Timber frame

Alejandro Aravena was awarded the Pritzker Prize for his contribution towards social housing in this global housing crisis. I know that my intention was not to cover social housing in this book, but his scheme caught my attention because of the unique offer to occupants to be able to expand their spaces within the given structure. Also, many housing schemes reserve some proportion to social or affordable housing so as to achieve a social mix within the community (Fig. 2.11).

The Villa Verde project provides worker housing for the forestry company Arauco, and it decided to apply the same strategy of incremental construction with higher construction and finish standards. On the ground floor, this base build comprises a small, shared space combining kitchen, dining and living room, plus a bathroom and outside laundry space. On the first floor, there are two bedrooms and an additional bathroom. The 57 m^2 houses are structured so that they can be readily expanded to a final area of 85 m^2. In its initial phase, the development comprises a terraced typology with each house separated room its neighbour by an undeveloped void. The pitched roofs form a continuous covering over this space, and at first floor, level joists are included that span to the adjacent party wall. Prior to further

Fig. 2.11 Villa Verde. Architects: Alejandro Aravena, Cristián Martínez of Elemental. *Courtesy* The architects

development, these covered areas provide flexible usable spaces. The plan configuration allows expansion of the accommodation into these spaces to provide a separate ground floor dining and living room and at first floor two additional bedrooms. As the circulation and the more complex building services are housed within the core structure, the residents are able to complete the expansion works themselves. Figure 2.12 shows plan layouts of the two floors with the ground floor having the optional use of the additional space increasing dining and living space. The first floor would have two bedrooms and a bathroom. The cost of the additional space use would be down to the occupant.

Floor area is 82 m^2.

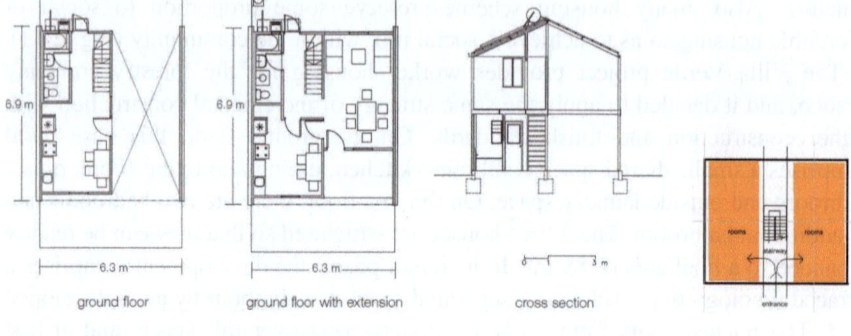

Fig. 2.12 Villa Verde plans and section

TECHNICAL

The buildings are constructed as timber frames, supported on concrete foundations, roofed in zinc and clad internally in 10-mm gypsum board and externally in 6-mm fibre cement board.

2.3.4 Abode Development, Great Kneighton, Trumpington, Cambridge. UK. 2014

Brick and block

Designed by Proctor and Matthews Architects, Housing Design Award 2012. Built on the former Clay Farm site, Abode at Great Kneighton is a key part of a major new housing and mixed-use development in South Cambridge. The houses sit within private-walled gardens and generously planted shared spaces and consist of 306 homes ranging from studio apartments to five-bedroom houses. Although the buildings feature a variety of different forms and cladding materials, London-based Architects chose to add alternating dark and light bricks at the entrances to help unify the scheme Floor area is 173 m^2 (Figs. 2.13 and 2.14).

These houses primarily serve the commuters to Cambridge and hence the importance of a bike store, which is becoming a regular fixture in any development that serves a town centre, and Cambridge is known as a cycling city. There are oddities compared with the more traditional housing schemes being a compact vertical three-storey four bed house as opposed to the normal two-storey ones. The end result is a rather soulless, stark impression. Hence, there is an integral garage, study and a bedroom plus en suite on the ground floor. The first floor has the main living, dining and kitchen space plus a double bedroom and separate bathroom as the toilet needs to be accessible to guests. Accepting that the occupants will have their own idea of furniture layout, the living area plan seems slightly odd

Fig. 2.13 Exterior of Abode development. *Courtesy* Proctor and Matthews. 2014

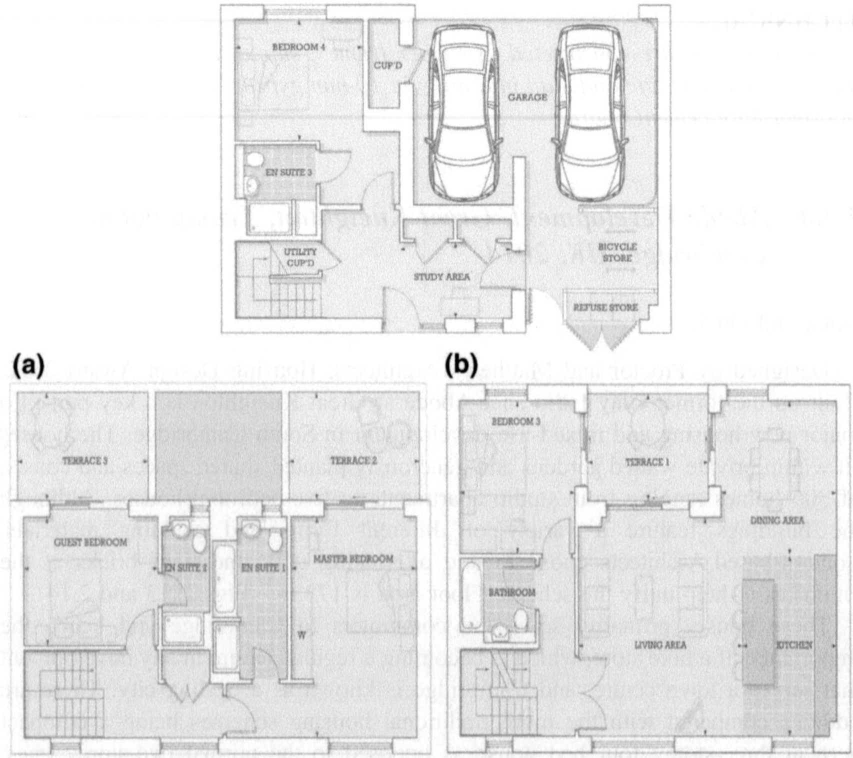

Fig. 2.14 Plan ground floor. Note bedrooms on each floor. **a** Second floor plan, **b** first floor plan

with regard to the television. It is not well sighted for viewers and seems to partially block access to the kitchen. Also, hi-fi equipment can take up some space and I do not see it working in this position. It may be better to position the TV against the left-hand wall. Kitchens work much better with a ‘C’ shape unit plan rather than the commonly favoured straight run, simply because it reduces circulation. So my conclusion is that the kitchen would work much better where the dining table is positioned because the walls are there (no need for two patio doors on the terrace), and that will allow better flexibility in planning ‘moveable’ furniture (kitchen is fixed) such as lounge, dining and media console. I visited the three-bedroomed terraced house types and was appalled to find that a total lack of storage was one of the main complaints of the resident I interviewed. I also found it curious that the presence of a huge column adjacent to the lounge area concealed a fresh air ventilation system, which was necessary due to the building being built as an airtight example for energy conservation. It is surprising that no utility space is provided. Some of these houses sell for £850,000, which reflects upon the successful growth of the hi-tech industries in Cambridge whose employees can afford such prices.

TECHNICAL

A limited palette of materials was used to accentuate this transition of spaces, with the dominant materials of Ibstock brick and perforated aluminium at the entrance gradually replaced by timber as the housing gets less dense. Windows are Velfac 200 series timber frame and aluminium sash Ral 7003. Beam and block ground floor structure with upper floor of Posi-joist timber.

2.3.5 Accordia Development, Four Bed Three-Storey House, Cambridge, UK

Brick and block as well as pre-cast concrete floors

Accordia was designed by Feilden Clegg Bradley architects working in collaboration with Maccreanor Lavington and Alison Brooks Architects. Another example of a soulless is prison camp kind of ambience. This substantial residential quarter in Cambridge is widely regarded as having set a whole new benchmark for large-scale housing in the UK and was the first housing project to win the RIBA Stirling Prize in 2008. Accordia comprised a mix of houses, apartments and duplexes, including 30% affordable mixed-tenure accommodation evenly split between housing for rent and low-cost housing for sale. This kind of development where the houses are interwoven with landscaped streets and pathways makes for identifying individual houses rather difficult. This is not a criticism but rather pays compliment to the integrational nature of its form and landscape whilst distinguishing between public and private space. The design replaces traditional gardens with a variety of private open spaces such as courtyards, roof terraces and large balconies. These three-storey houses emphasise the public amenity by creating communal first-floor space and included 9-m-high chimneys feeding all apartments. But to me, they appear soulless and lacking any humanity (Figs. 2.15 and 2.16).

The internal planning is very good with sufficient space provided for decent living. The ground floor plan shows us an integrated single garage with access into a hallway. This is the reception floor with cloaks, utility, and open-plan kitchen, dining and living space. This open-plan arrangement I criticise later in the book. There is also a unique access to two terraces; one on the first floor and one off the bedrooms on the second floor.

TECHNICAL

Materials:

The external materials for the housing comprises mostly of bricks selected to match closely the traditional Cambridge Gault clay bricks, with the apartments constructed from copper and green oak. Many elements of the buildings were fabricated off-site—increasing speed of construction, reducing waste and improving environmental performance. As they age gracefully, the stock bricks, composite timber and aluminium windows, and all copper and untreated hardwood will require minimal maintenance.

Fig. 2.15 Three-storey 'chimney' four bed houses of Accordia

Fig. 2.16 Plans from left: ground, first, second, roof. Section above

2.3.6 *Zed Factory, Hastings, UK—Five Terraced Houses 2016*

Timber frame

This development of five zero carbon homes which can provide four bedrooms and a single garage is located near Hastings castle. Perched high up between Hastings Old Town and the Town Centre, it sits in a residential setting within the Old Town Conservation Area. Zed factory has worked up a design which both delivers in design quality and environmental performance, providing an opportunity to transform this derelict site into a historically important project for the city to coincide with the 950th anniversary celebrations of 1066. The design is of a staggered terrace of five houses with the top floor of each home being a flexible open-plan space intended for a range of activities depending on the occupants' lifestyles. This could be a studio, office, living room, games room or workshop for example. This is an encouraging trend that is responding to the modern needs of a family. The brightly painted canopy/walls are reminiscent of the Oxley Woods project in Sect. 2.2.1, whilst the pitched roofs pay vague homage to the surrounding Edwardian properties. The designers have taken advantage of the sloping land and orientation to maximise on the solar gain. They have also incorporated garages to each property enabling terraced gardens to be made on top of them (Figs. 2.17, 2.18 and 2.19).

The open-planned space incorporating kitchen, dining and living seems to be the fashion as we have seen with past case studies, but I maintain that this creates conflicting activities which will in the end make the occupants unhappy with living

Fig. 2.17 Exterior view of the housing. *Courtesy* ZedFactory. *Photo* Duan Fu

Fig. 2.18 Plans

Fig. 2.19 Elevations of the houses

here. There is no utility area. The entrance hall is far too small with no indication of where hats and coats are to be kept. The bedrooms have good spaces with the option of creating an en suite bedroom on the top floor. There is a small cramped study on the first floor. Total internal floor area is 139 m^2 (Fig. 2.18).

TECHNICAL

The construction is a heavyweight timber frame, allowing for abundant insulation and a good level of thermal mass in the form of screeded floors to each story. The external walls are insulated with a highly sustainable natural wood fibre product, to achieve an overall U-value of 0.14.

Windows are timber frame aluminium clad triple glazed units manufactured by Velfac, and the folding sliding doors on the ground floor by Sunflex. The South roof is entirely made up of the ZEDroof solar panel roofing system, and the North is a standing seam Zinc. Gutters are all metal, and timber cladding is from locally grown sweet chestnut. Heating and ventilation are provided by a combined system, incorporating an air source heat pump and MVHR unit. Heating is distributed by underfloor electric heating coils embedded in the floor screed.

2.3.7 *Orchard Close, Almondsbury, Bristol, UK*

Architects Stride Treglowan—2013

Development of four terraced three-storey four-bed houses with single garage. Masonry cavity.

This project located to the north of Bristol is aimed at the high end of the residential market. The houses are sustainable and lifestyle-friendly. They exploit a southerly orientation, and the houses radiate in an arc to capture the sun.

Like a piece of origami, the outside and inside walls, roof and floor are seemingly made from one continuous material and 'wrap' around the building to create a strong inside–outside experience, blurring the transition between house and garden. The south-facing elevations are highly glazed to capture low-angle winter sun. Timber slats and deep reveals reduce solar gain in the summer, and roof lights at the top of a three-storey height space create a passive stack cooling effect. Internal floor area is 234 m^2 (Figs. 2.20, 2.21, 2.22 and 2.23).

Fig. 2.20 Front of house. *Photographer* Mike Gove. Rear photo courtesy Savills

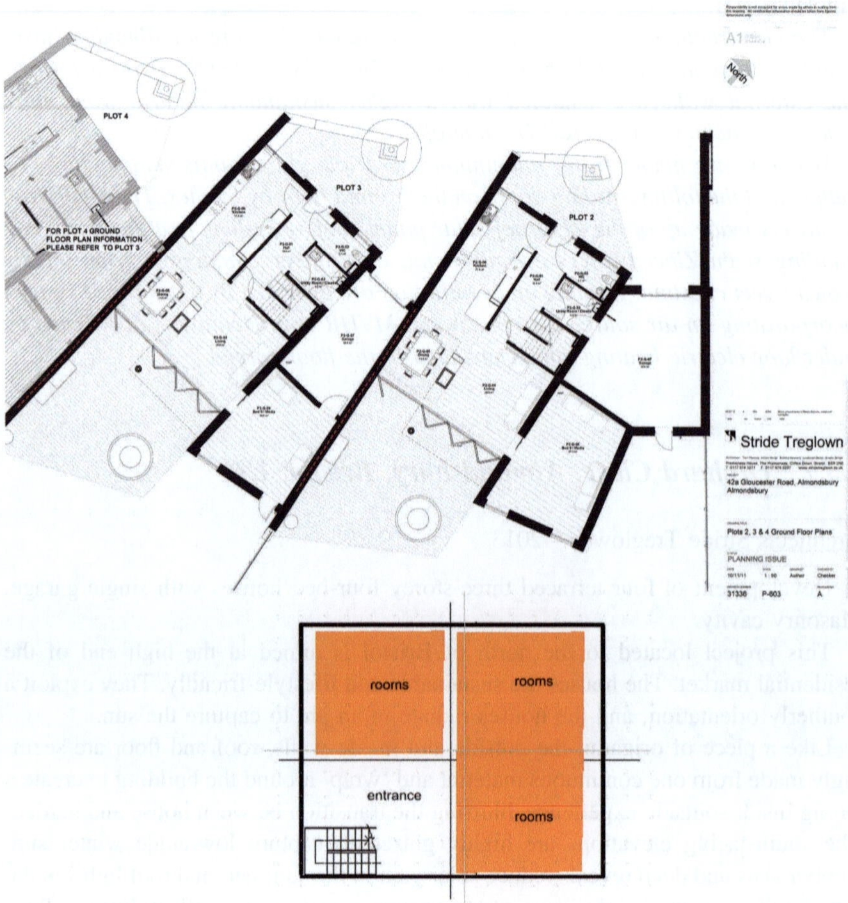

Fig. 2.21 Ground floor plan

Unashamedly aimed at the luxury end of the market, the architects are to be congratulated for providing expressive exterior forms that honour the weather and solar orientation. They are larger houses than any of the case studies in this book, and so it follows that there is ample space and storage provided. The 'extra' space compared with smaller houses must be due to the snug at the rear of the garage, and the triple height space above the dining area. It is good to see that bedrooms are of good size and that a dogleg stair is incorporated instead of the wretched straight stair. If luxury was the aim, I would have expected better coat and shoe provision in the entrance hall and more space in the utility room. Also, considering the increase in the use of such devices as laptops, iPads and mobile phones used by ALL members of the family a shared studio space is required rather than just naming the fifth bedroom as an office which is too small anyway. These devices are more used in the larger communal areas than tucked away in a spare bedroom. My only other quibble is that some windows are floor to ceiling and that is one of my 'no's' in the Introduction.

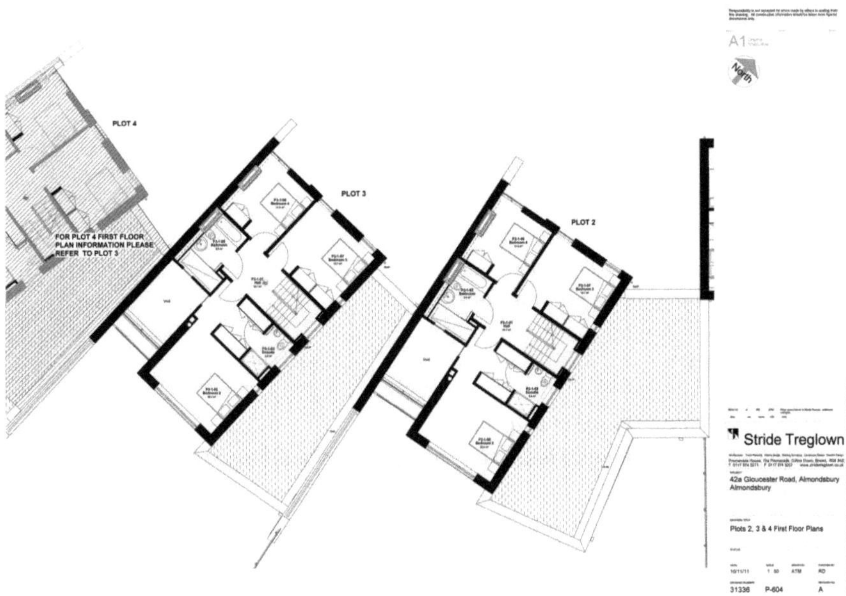

Fig. 2.22 First floor plan

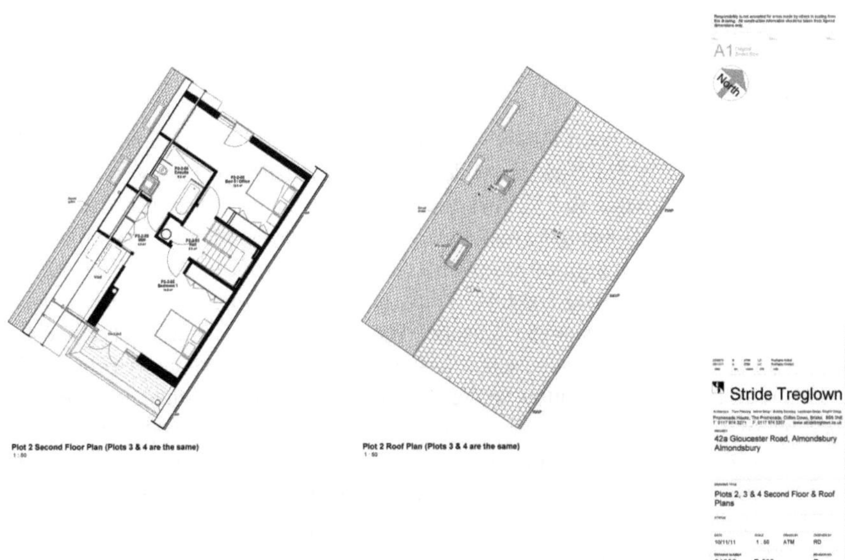

Fig. 2.23 Second floor plan

TECHNICAL

Walls:

- *Masonry cavity-wall construction*

Services:

- *Conventional gas combi boiler*
- *Underfloor heating to ground floor*

External finishes:

- *Ibstock tile brick (vertical wall tiling)*
- *Natural fibre compound cladding (vertical wall cladding made from rice husk)*
- *Render*
- *Tiled roof (main roof)/sedum roof (garage/snug roof)*
- *Aluminium windows*

2.4 Semi-detached Houses

2.4.1 Isle of Wight, UK, Cameron Close, Passivhaus Development 2009+

Brick and block

Cameron Close is a good Passivhaus[3] four beds no garage development in fresh water on the Isle of Wight. The mixed development is made up entirely of affordable units, including 16 houses and 12 sheltered apartments on a 1.5-acre site. The development for Southern Housing Group aims to provide homes that are low on energy consumption yet comfortable and healthy for the residents to live in. All the houses have front and back gardens and the apartments have generous, landscaped ‘kitchen gardens’ to ensure plenty of private amenity space. The varying size of the houses and the flexibility of their plan will encourage a sustainable, cohesive community (Fig. 2.24).

This is an interesting variation on the more common semi-detached arrangement, whereby two houses mirror each other. Here we have two joined but different houses, the term of which I prefer to call ‘conjoined’. The left-hand narrow house H1 with side entrance is on three floors with a bedroom on each floor and has one double bedroom, two study bedrooms and one single bedroom. It also has two bathrooms and one separate WC (Fig. 2.25).

[3]Passivhaus: is a rigorous, voluntary standard for energy efficiency in a building, reducing its ecological footprint. It results in ultra-low energy buildings that require little energy for space heating or cooling.

Fig. 2.24 Conjoined houses in Cameron Close, Fresh water, Isle of Wight. *Architects* PCKO. *Courtesy* The architects

Fig. 2.25 Plans of two Cameron Close houses

The wider house H2 with front entrance is on two floors and has one double bedroom, one study bedroom and a shared single study bedroom with two single beds. Curiously, only one single bedroom in H1 has a balcony. The greater width of H2 is only experienced in the kitchen/dining room; otherwise, elsewhere in the house it has the same narrow experience as H1. This deception is felt when the house is viewed from the front, leading one to think that H2 is the bigger house, but this is not the case. There are no letterboxes in the front doors as this would break the airtight seal of the houses. Mailboxes are fixed outside each house. There is still a lack of storage space, and it lacks a utility area.

H1 house—floor area 114.75 m^2
H2 house—floor area 109 m^2

TECHNICAL

The H+H thin-joint system, which the houses are built by combining the Celcon Plus, Jumbo Blok or MultiPlate ranges of high quality, accurately dimensioned aircrete blocks with a specially developed thin-joint mortar (called Celfix). Thin-joint blockwork enables walls to be built very quickly to storey height without having to wait the conventional 24 h for the mortar to set before further loading can be applied. Fully adopted as the preferred method of wall construction throughout most of Europe, this innovative and award-winning system is now firmly established within the UK. The thin-joint system can also be used successfully for commercial projects. The external solid walls are insulated with min 200-mm PIR insulation boards and finished externally in white render, a fairly common aesthetic for Isle of Wight homes. Another innovative aspect was the use of retrofit joist hangers for the internal floors. This avoided the need to cut blocks or penetrate the walls. This removed any concerns with regard to air tightness performance and sound transfer.

TECHNICAL PERFORMANCE

Fabric U-Values
Ground floor—average 0.1 W/m^2 K: beam and block floor; 250 mm PIR insulation; 125 mm power floated reinforced concrete screed. Foundation walls to splash course and brick and block with cavity filled with XPS insulation.
Roof—average 0.8 W/m^2 K: including either 600 mm Rockwool (lofted areas) or 250 mm PIR (flat roofs).
Windows—0.96 W/m^2 K: Passivhaus certified, triple glazed; doors Passivhaus certified, solid.

External walls—0.8 W/m^2 K (houses), 0.11 W/m^2 K (apartments): acrylic render system; 230 mm phenolic insulation (150 mm to apartments); 200 mm solid thin-jointed aircrete 3.6 N/mm^2; plasterboard on battens.

Airtightness

All units achieved below 0.6 air changes per hour at 50 Pa (n50 tests)
Sound testing
Separating walls: two leaves 100-mm thin-jointed aircrete with 100-mm full fill insulated cavity
Airborne, separating walls for houses and apartments average 57Db D DnT, w +Ctr
Separating floors: E-WM-14 with 200 mm PCC planks, 75 mm screed
Airborne, separating floor to apartments average: 50 dB DnT, w +Ctr
Impact, separating floors to apartments average: 50Db L'nT, w.

2.4.2 Letchworth Three Bed Three-Storey Houses Plus Single Garage, UK Architects Higgs and Young. 2012

Total internal floor area 105 m^2 excluding garage.

This fine pair of houses with its Rietveld[4] inspired exterior modelling displays variable elevations that continuously surprise the viewer. As the architects say (Figs. 2.26 and 2.27): *Two houses completed in 2012 for North Herts Homes selected from 2007 competition Tomorrow's Garden City as code 4 housing to reduce carbon emissions, reduce and recycle waste and conserve water, thus enabling the vision of the garden city to be realised - enabling people to live in communities which are both closely integrated and in harmony with nature.*

The disappointing aspect about this scheme, although the architects were working to a tight density brief, is the planning and lack of interior provision despite the floor area being 105 m^2. The ground floor is very tight with a small hall and a WC almost in your face on entry. The lounge appears small having to share with the dining area and kitchen; a common undesirable trend. The scheme contains all of my four No's as well as lacking any utility or study area. The integrated garage seems small and lacks direct internal access but sits well with the architecture of the houses (Figs. 2.28 and 2.29).

[4]Gerrit Rietveld. 1888–1964 was a Dutch furniture designer and architect. One of the principal members of the Dutch artistic movement called De Stijl, Rietveld is famous for his Red and Blue Chair and for the Rietveld Schröder House.

Fig. 2.26 Exterior view of one of the pair. *Photo* Morley von Sternberg

TECHNICAL

Dwellings have a roof suitable for installation of solar thermal heating and photovoltaic panels. The 30° sloping roof has brown sedum roof finish which is capable of being retrofitted with additional photovoltaic solar panels.
Specification: Code for sustainable homes level 4, timber frame, rendered Hempcrete walls.
Monodraught passive ventilation.
Aluminium solar shading louvres.
Engineering brick base.

Fig. 2.27 Letchworth elevations

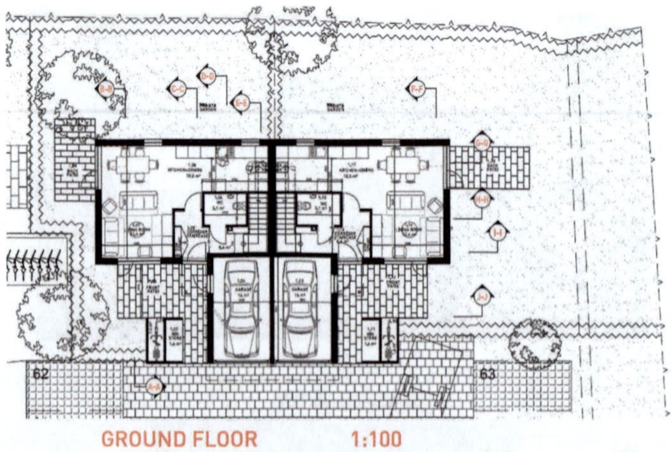

Fig. 2.28 Letchworth ground floor plan

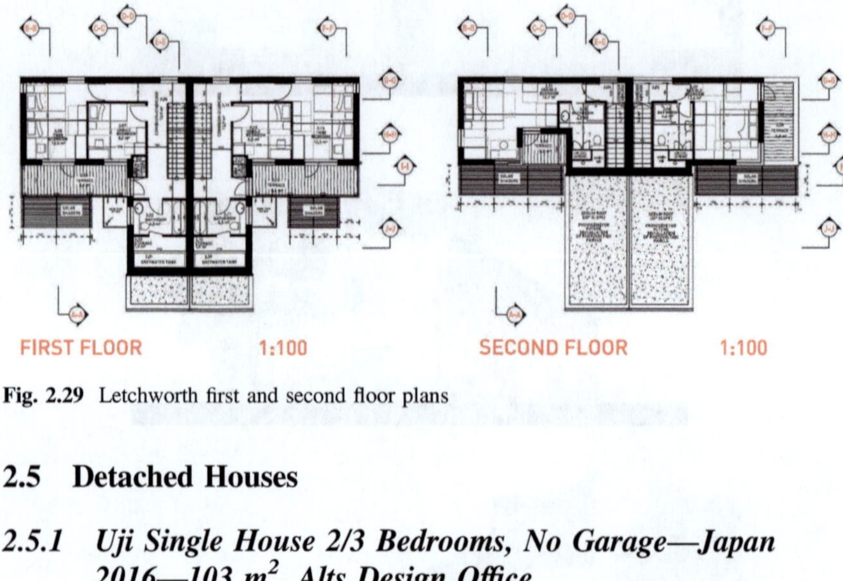

Fig. 2.29 Letchworth first and second floor plans

2.5 Detached Houses

2.5.1 Uji Single House 2/3 Bedrooms, No Garage—Japan 2016—103 m^2. Alts Design Office

This house is located in a housing estate of Uji city and surrounded with old houses, which have been rebuilt. Privacy for the occupants has been the prime consideration. Timber frame construction (Figs. 2.30 and 2.31).

The Japanese are masters of making use of every square millimetre, and this house is no exception. It is a private house that follows the severe rectangular grid, which the Japanese find difficulty getting away from. Yet, there are interesting elements that we could learn from. With compact spaces, there is still an open-plan

Fig. 2.30 Uji House exterior. Two/three bedrooms. Kyoto, Japan. Alts design office 2016. *Photographer* Yuta Yamada

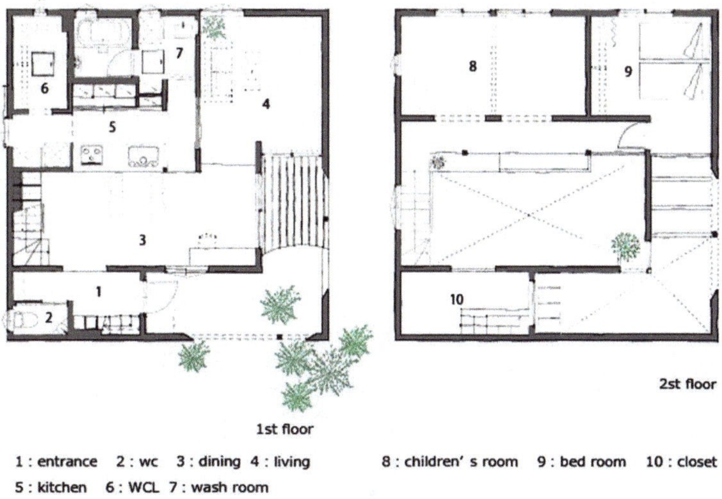

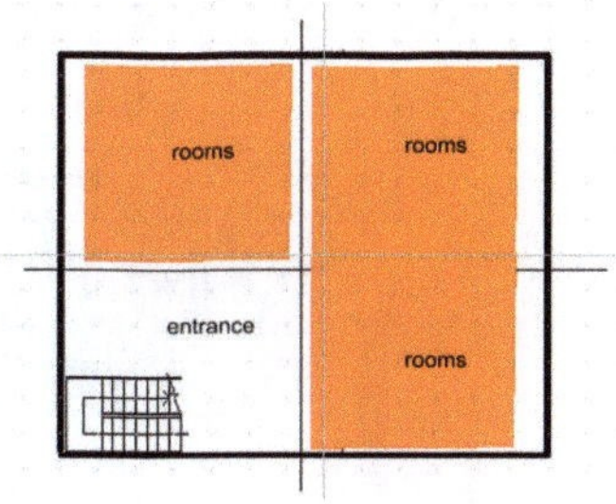

Fig. 2.31 Uji House plans

feeling between the kitchen and dining area which then flows into the living space without there being 'rooms' as such. The double height space above also helps convey spaciousness. The children's bedroom can be divided into two when their age requires it. Main bedroom under 12 m^2.

2.5.2 *The Tin House Holiday Home by Rural Design 2016*

Rural Design has completed a shed-like one-bedroomed holiday home with no garage, on Scotland's Isle of Skye, featuring corrugated aluminium walls that reference local agricultural buildings. The aluminium surfaces help to protect the building from the fierce storms that regularly batter the island, with minimal window openings. The building consists of a double bedroom, a living area, kitchen and bathroom. Internally, the architects aimed to create 'a character that is simultaneously modern and rustic' which was achieved by combining simple materials such as a concrete floor, plywood cabinetry and timber wall panelling. Again another example of compact space is well planned. Rather than design a 'box', the pitched roof provides more height by using the angle of the roof to compensate for the rigid rectangular planning. It also, as stated above, relates to the local agricultural buildings on the island. Bedrooms' estimated area is 15 m^2 (Figs. 2.32, 2.33 and 2.34).

Fig. 2.32 Exterior of the Tin House, Isle of Skye, Scotland. Rural Design 2016. *Photographer* David Barbour

Fig. 2.33 Living space

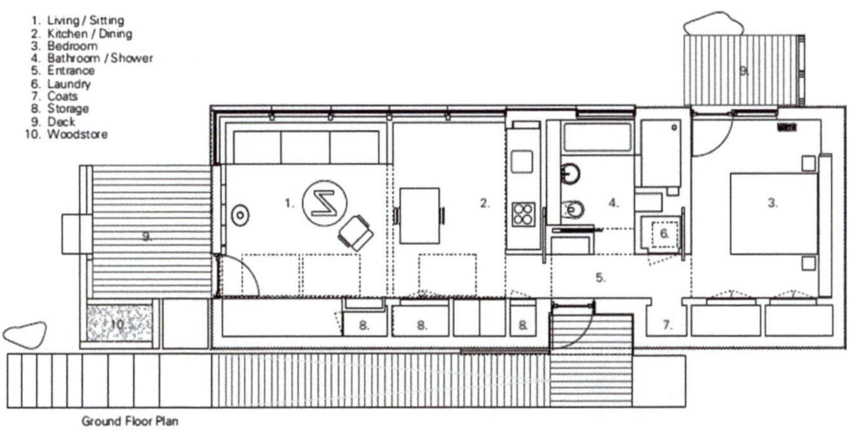

Fig. 2.34 Tin House plan

TECHNICAL

The project combined simple materials such as a concrete floor, plywood cabinetry and timber wall panelling. Recycled materials were employed wherever possible, including for the sliding pocket doors that can be used to close off the living space, bathroom and bedroom. These doors feature circular cut-outs instead of unnecessary handles.

2.5.3 Three Bed Single House No Garage in Balsthal, Switzerland 2007–2012. Pascal Flammer Architects

This timber frame house is about different ways of perceiving the landscape surrounding it. It is in isolation from any community. There are two principal floors; one set 750 mm below the earth, one 1500 mm above. The ground floor consists of one single family room with a noticeably low horizontal ceiling. In this space, there is a physical connection with the nature outside the continuous windows. The space above is the inverse. This floor is divided into four equal rooms with 6 m height to the ridge (Figs. 2.35, 2.36 and 2.37).

Area 192 m^2.

Fig. 2.35 House at Balsthal, Switzerland by Pascal Flammer. 2014

Fig. 2.36 Balsthal house Ground floor interior

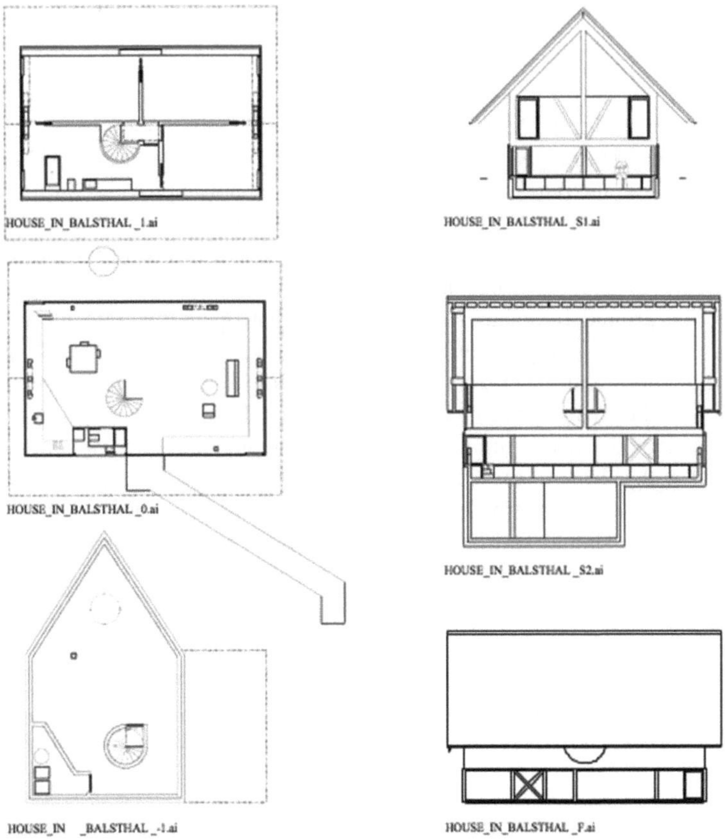

Fig. 2.37 Drawings of the Balsthal House. Top left is first floor three × bedrooms plus bathroom. Plan below that is the Gd floor open-planned general space with kitchen

There is a beauty and simplicity in the sense of space, the dominance of wood and the relationship to the external environment. It is such a relief to see a departure from the more common house designs as we have seen so far in this book. I especially like the pocket doors to the bedrooms—a method that I shall be using later in the book. The large open space on the ground floor is undedicated in much the same way the continuous panoramic views are through the windows. The continuous storage benching is again undeclared and open for grabs by the activity it serves. Brilliant and brave at the same time.

2.5.4 Hanse Haus, Germany—Prefabricated Self Build

Timber frame

The image in Fig. 2.38 is a three bed house with a total floor area is 170.60 m^2, that uses a 'Fabric first' system which is used to build certified Passive Houses. As can be seen from the plans in Fig. 2.39, this German house provides far more space than the equivalent in the UK and it is the best designed plan from all the case studies in this book. It is good to see a winder staircase instead of the straight stair. The exterior is uncomplicated and strongly defined but a little conservative in design. The ground floor has a generous hall that opens out into the combined kitchen, dining and living space, as well as providing access into the WC, a decent size utility room and a generous study. The offer of an optional bay windowed extension to the living space creates the only modelling on the exterior.

The combined kitchen, dining and living space amounts to an area of 44 m^2 as compared with 22 m^2 in the Oxley Woods house, although admittedly the Oxley Woods house is a much smaller three bed house. However, my point is that the communal areas should still be of a constant size irrespective of how many bedrooms there are. I quote them from the company's Website:

> *Building a house is more than merely creating new living space; it's also about putting ideas into practice to give quality of life and sense of freedom within four walls. The Variant 25-166 is a house where the whole family can find space for their own ideas.*
>
> *While the living and dining room with its adjacent open kitchen is the ideal setting for an active family life, the top floor with its two children's rooms, master bedroom with dressing*

Fig. 2.38 Hanse Haus exterior. *Courtesy* Hanse Haus. 2016

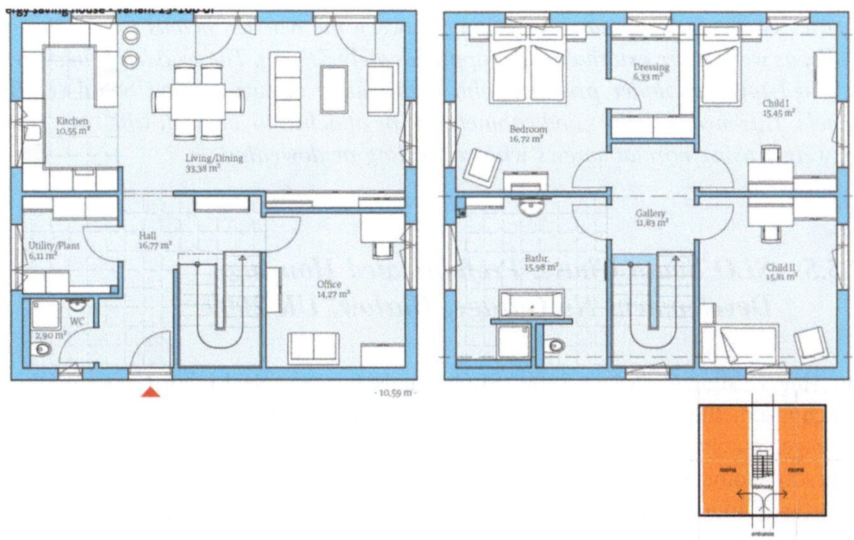

Fig. 2.39 Plans of ground floor on left and first floor on right

room and family bathroom offers space to withdraw for some peace and quiet. Ingenious storage spaces in the generous entrance area prevent overspill of coats and shoes. A bay window on the eaves wall of the house expands the living room on the ground floor by around 4.5 m^2. This gives a range of new possibilities, for example in the arrangement of the dining area.

TECHNICAL

Solar energy: Environmentally friendly and free

As solar energy makes up the primary contribution to the heating, it is important that lots of solar rays fall on the house. If it were to become Passive then this is best done by having the house face the south.

Extremely high thermal insulation

Thanks to this extremely high level of thermal insulation and airtight building shell, almost no heat is lost through the walls, roof or windows. The heating energy required by the house is almost completely made up by solar energy. The rest of the warmth is provided by the residents themselves, with each person giving out around 80 W of energy per hour into their surroundings.

Windproof building shell and modern ventilation technology

High demands are placed on the tightness of a passive house. It is very important to make sure that no cool outside air can get in and take the valuable heat out of the house. For this reason, all the joints in the external walls are interlocking such as tongue and groove. The roof and the windows and doors must be carefully sealed.

Walls

Hanse Haus thermo-efficiency walls combine a high-strength construction method with excellent thermal insulation. In wood frame construction, Hanse Haus focuses particularly on narrow spacing between the wooden beams on the inside walls, as well as the external walls (approximately 31 cm). The wooden framework is glued together under pressure with high-quality oriented strand board *(OSB) panels. This means shelves and cabinets can be attached to any wall with over 100-kg weight using normal screws without drilling or dowelling.*

2.5.5 SLO South Chase, Prefabricated Housing Development No Garage, Harlow, UK 2009

Housing Design Award in 2009. SLO Stands for 'Simple Living Opportunities' which is a method of accelerating customisation to achieve a time-frame demanded by eager purchasers. This had not always worked out but the desired result has produced some very finely designed housing. This example is a prefabricated 'courtyard' four bed house which produces a 'C' shape plan that is far more inclusive of space and privacy that can ever be offered by a 'box' (Figs. 2.40 and 2.41).

Fig. 2.40 Exterior of one South Chase House type. *Courtesy* Proctor and Matthews architects

Fig. 2.41 Ground- and first-floor plans

The whole development is refreshingly varied contextually by form and materials. A covered entrance space and courtyard have direct connection with the kitchen whilst the living room enjoys a dual aspect having glazed full-height doors to both garden and courtyard. Set at the 'heart' of all the homes is the stair and kitchen, considered together as a piece of bespoke furniture. The utility machines are under the stairs which is similar to my own design proposal in Chap. 5. The entrance is protective and fairly private aided by the covered courtyard. There is no hall as such before you realise that you are in the dining area. As with the majority of schemes in this book, the term 'room' has become defunct regarding the 'reception' spaces of the house because of the open-plan nature of the planning. The toilet is, however, in close proximity, as it neatly forms part of the central block housing the washing machine and drier as well as the staircase. Unfortunately, this is a straight stair which is the most dangerous compared with the dogleg stair. The kitchen is partially concealed from the lounge area, which is given a decent amount of space as well as being a good rectangular proportion away from the 'railway carriage' syndrome with so many modern housing. For example, it is rare to be able to fit two three seater settees at right angles to each other as is possible with this house. This lounge area rectangle is very close to the golden section thus emanating that ancient aura! What completes the openness of the design is the amount of windows (some down to the floor) that accept a good amount of daylight from three directions. A study area is missing from these plans. Most housing concentrates, because of siting, on main views to the front and rear, hence two directions.

The first floor is beautifully planned efficiently and sensibly. I have noted, however, that the master bedroom is one square metre below my recommended minimum of 14 m^2. The bathrooms are undersized especially the en suite where washing at the basin is almost impossible without bruising elbows against the wall. My only other point to make, and this applies to most bathrooms, is that there is never enough space to store the incredible amount of toiletries that ladies seem to use. I am agog at the profuse amount on display in supermarkets! If both partners of a house have to leave for work the same time this suggests two basins in the en suite to ease panic congestion. This is such a much-needed requirement, but architects have fallen into the 'mean' trap. Total floor area is 92 m^2.

TECHNICAL

The house utilises the future form building system, a monocoque steel construction. Modules arrive on site sealed and weather tight with a full internal fit-out. They are then stacked, roofed in clay tiles and faced in classic Essex white render before prefabricated porches, canopies and balconies complete the exterior elevations.

2.5.6 Compact Two Bed No Garage Karts House, Slovenia, 2014. Dekleva Gregorič Architect

Concrete and stone

I have chosen this small house for its design ingenuity and minimalist use of space. The design of this small compact stony house corresponds to the needs of a young family and simultaneously follows current technological principles and the tradition of small, compact, stony and almost windowless houses. The ground floor operates mostly as a public or semi-public space with grand landscape views, whereas on the other hand the upper floor stands very private, with sky views only. The space is divided with two inserted wooden volumes which contain a kitchen with dining or bathroom on the ground floor, and a master bedroom and children's room on the upper floor. The house in a house concept makes one feel like sleeping in one's own house and not a room. The bridge connecting both houses acts as a playroom. The house has three large square windows which look out on views towards a hilltop church in Italy to the west and a forest to the south (Figs. 2.42, 2.43, 2.44, 2.45 and 2.46).

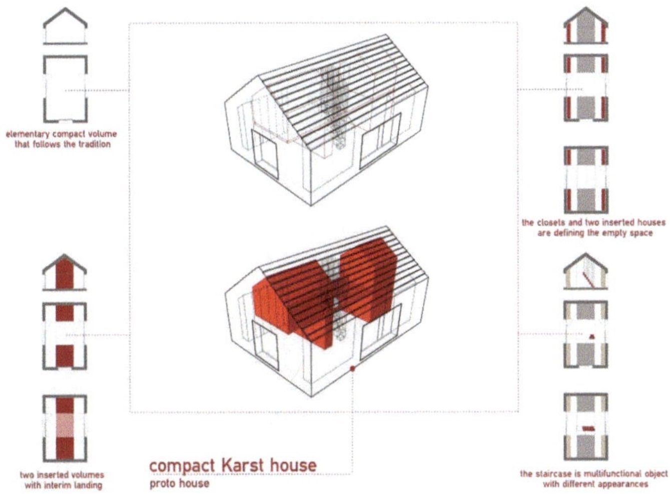

Fig. 2.42 Karst House schematic drawings

Fig. 2.43 Exterior. Photos Janez Marolt

Fig. 2.44 First floor view

Fig. 2.45 Ground floor living space with stairs

The skill in the planning is the containment on both floors of one central linear block holding all activities leaving the side channels for circulation access to storage lining the walls on the ground floor. The staircase doubles up as storage for books under and is structurally tied into the heavy house form by slender metal rods.

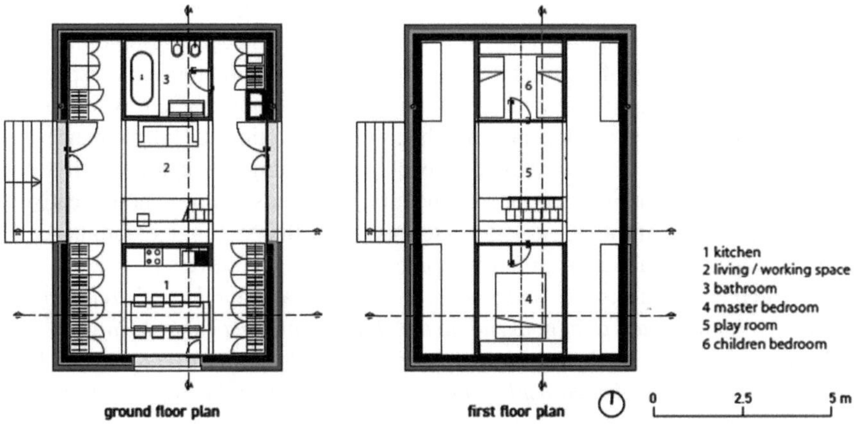

Fig. 2.46 Plans of Karst House

TECHNICAL

The redefinition of a traditional stony Karst roof, with its texture, colour, material and its steep inclination is executed as a contemporary concrete interpretation. A materially inseparable connection between the facade and the roof is a key allusion to the image of the traditional Karst village. The stone, set into the concrete during the casting process, gives the walls the appearance of having a solid masonry construction—this 'low tech' technique previously employed by Aljoša's architect father in the late 1970s in the Karst region. The ceiling is the concrete slab of the roof construction—the formwork was made from wooden planks. All interior dry walls are made of three layer spruce plywood panels, oiled with transparent oil. The horizontal slab is spruce cross-aminated timber.

2.5.7 Murphy House at Hart St, Edinburgh, UK. 2014. Three-Bedrooms and Garage

Steel frame

Designed by Richard Murphy with Gareth Jones, James Falconer, Tersius Maass.

Floor area 165 m^2.

Awards:

2015 Saltire Society Housing Design Award
2016 RIBA/Channel Four House of the Year 2016
2016 RIBA Award
2016 Edinburgh Architectural Association—Building of the Year 2016
2016 Civic Trust National Award
2016 Architects' Journal UK House of the Year

This house is not within my range of focus of mass housing as it is a bespoke solution for the architect himself but, as with some other projects, it deserves mention for the sheer creativity of maximum use of minimal site and 3D form. It is especially notable for consisting of a wealth of moving parts like some magic box that opens up. There are sliding doors, opening fanlights, sliding, pivoting and hinged panels that reveal windows, etc. It conforms to one of my teaching strategies which I call 'Mechanics of Operation', designed to encourage students to think more creatively about the potential of moving parts in interiors rather than accept the dominance of static architecture with its token moving parts of doors and drawers (Fig. 2.47).

The architect writes: *The house sets out to achieve a number of architectural ambitions. Firstly it acts as a 'bookend' to the above mentioned gable, hiding as much of it as possible. The elevational treatment continues the pattern set up by the Hart Street houses of an indented ashlar base, string courses and a significant cornice which is now terminated by becoming the roof edge of a dramatic sloping roof. The external form of the house is completed by a garage with a small roof terrace above.*

Fig. 2.47 Top left daytime exterior. Right night-time exterior. Below left rear view. *Photographer* Keith Hunter. *Courtesy* Richard Murphy

The house consists of three double bedrooms, integral garage, store/plant room, log store, wine store, utility room, hall, study, open-planned split level living space with kitchen and dining, terrace, one toilet, two toilets plus shower and one bath in the master bedroom. Looking at the plans, there is no classical framework that everything hangs on. There is the external structure and two staircases; one narrow one with access from basement right up to second floor. The wider staircase is just from ground to first floor. One could say that it is a planning mess but that is to deny the ingenuity of its 3D planning (Fig. 2.48).

TECHNICAL

The front façade continues the stonework pattern of the street façade concluding with the idea of 'an inhabited ruin'. Even when a building is completely new this is a theme which frequently reoccurs in our architecture when building in historic places. The 'new' architecture is layered with contemporary materials of glass block, steel, burnt timber and lead used in a tectonic way contrasting with the solidity of what might be a pre-existent ashlar stonework 'ruin' to either side. Within the ashlar is a pattern of tiny windows (which sit amongst the shelves of a giant staircase-bookcase inside) and these play on the ashlar stone construction

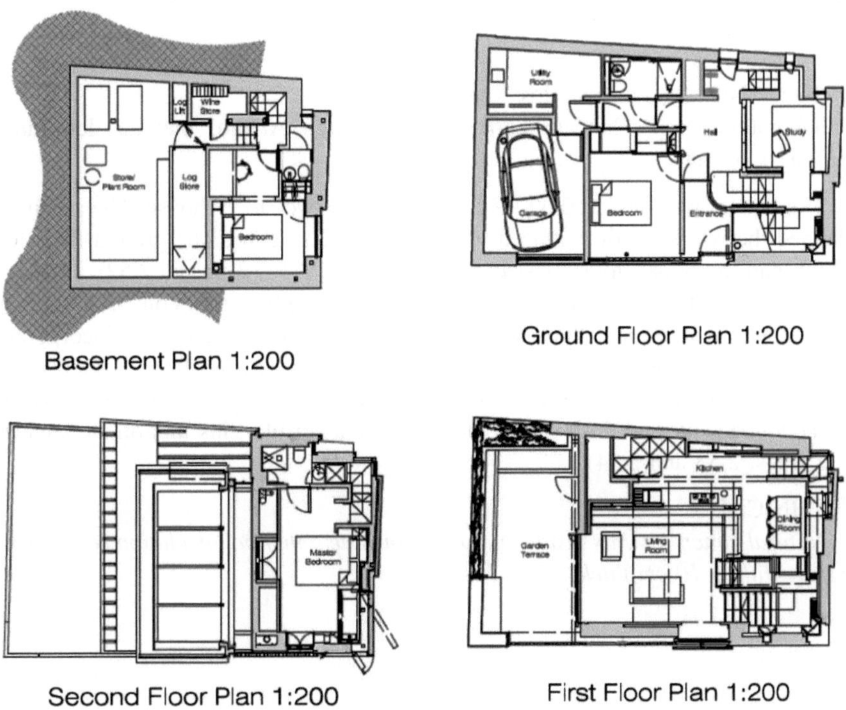

Fig. 2.48 Plans of the Murphy House with the street frontage south of the drawings

with corner windows reminiscent of coin-stones, but in the negative. As with the adjacent tenements, the ashlar turns to rubble at the rear where the whole elevation becomes a much freer composition. The roof made mostly of glass with inset photovoltaic cells is designed both to ensure daylight to the adjacent basement flat on Forth Street and also to act as a major collector of solar energy. Inside the roof are a number of insulated shutters, which are capable of closing when the roof is in net heat loss mode and opening when there is a net heat gain. Other energy innovations are a computerised internal air circulation system which draws air from the very top of the house to the semibasement to both counter the stack effect but also to store heat in a rock store placed in the solum for night-time heating. The main heating source for the house is a 150-m-deep ground source borehole connecting to a heat exchanger which feeds underfloor heating. Rainwater which follows a course of pools and waterfalls on the roof terrace finds it way to grey-water storage tanks in the basement and is then used to flush toilets and supply a sprinkler system. In winter, heat is extracted from the flue of a log burning stove to pre-heat hot water.

2.5.8 Three Bed Mami House No Garage, Matoshinos, Portugal, 2016

Architect: NOARQ, José Carlos Nunes de Oliveira.
Collaborators: arch. Gonçalo Ferreira; arch. Hugo Araujo. Internal floor area 162 m^2. Concrete.

The budget was only 100.000 € for a three bed two-storey house. The architect planned the whole house within a 9 m × 9 m × 6 m cube. The ground floor for living, and the upper floor for sleeping (Figs. 2.49, 2.50, 2.51 and 2.52).

The simplicity of this design is its strength although it does contain a straight stair, square corners and floor to ceiling windows. However, it does incorporate pocket sliding doors upstairs which I shall be using in my design proposal in Chap. 5. The ground floor core is conceived as a wooden block lined with shelving on the lounge side and flush storage doors and a peninsular sink unit on the kitchen side (Figs. 2.53 and 2.54).

TECHNICAL

Natural concrete walls inside and on the outside a thermal insulation component system with 8–10 cm thick.

Fig. 2.49 Mami House, Portugal. *Photographer* João Morgado

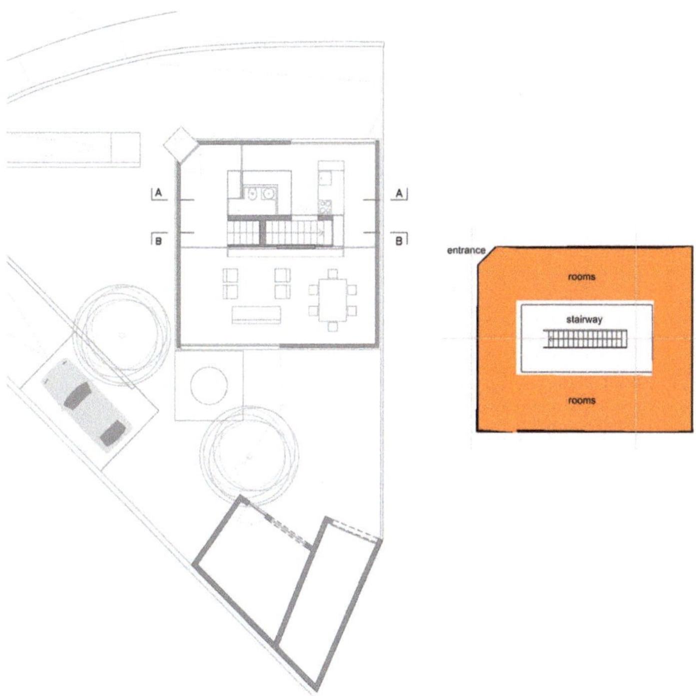

Fig. 2.50 Ground floor plan with entrance door in the corner

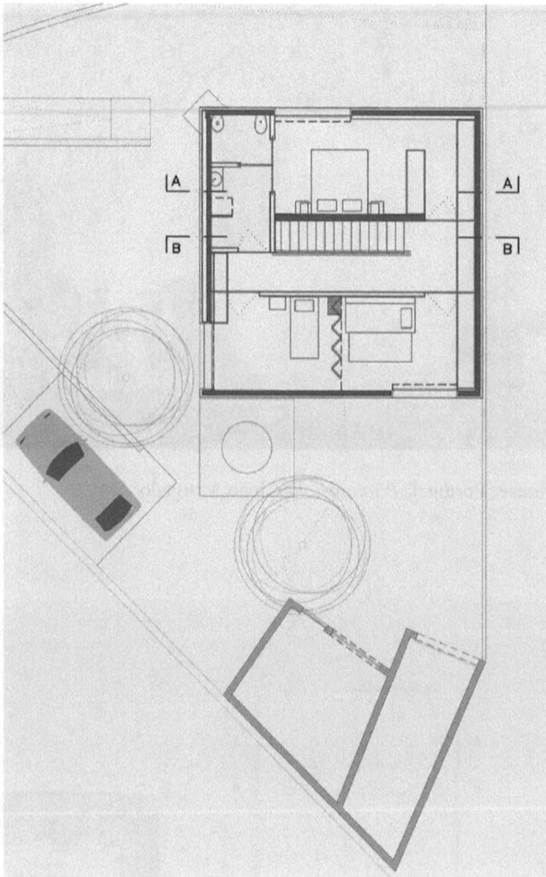

Fig. 2.51 First floor plan

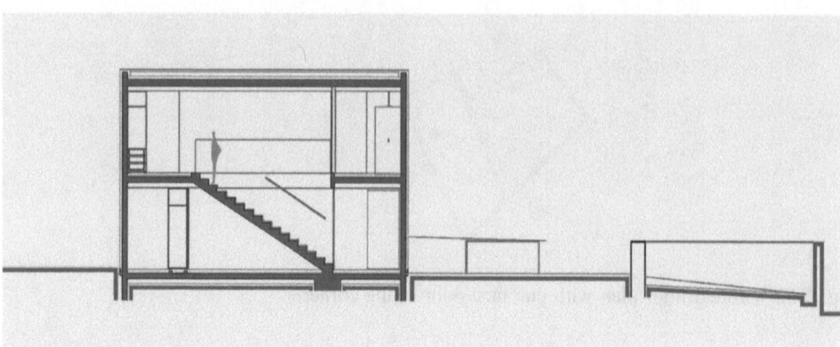

Fig. 2.52 Section

Fig. 2.53 Kitchen

Fig. 2.54 First floor sliding pocket doors to wash area

2.6 Cellular Designs

2.6.1 Element House—Guest House and Visitor Centre—New Mexico, 2014. MOS Architects

SIP's panels

Credit: Commissioned and built by the Museum of Outdoor Arts as the guest house for Star Axis an eleven-story sculptural observatory created by Charles Ross.

As I stated previously, not all the projects featured in this book, including this one, fall into any of my main categories. This is a study centre and therefore has a generous amount of open meeting spaces, a relaxed atmosphere and is at one with the surrounding natural environment. There is so much originality in the scattered layout, and hence, it follows organic principles. There are surprising angles and chimney shafts of light from above. The Element House is a guesthouse and visitor centre for Star Axis, a land-art project located in the foothills 120 miles east of Albuquerque, New Mexico (Fig. 2.55).

The building acts as a gallery and guesthouse for visitors and as a prototype for off-the-grid housing. The project was awarded the 2015 Award of Merit from the New York State AIA (Figs. 2.56 and 2.57).

TECHNICAL

Constructed through an aggregation of prefabricated modules, the building can be accessed from all sides. A decentralized field of solar chimney clerestories perforates the interior to provide indirect, natural light and consistent ventilation. The building's aluminium shingle skin acts as a heat sink, floating above a small airspace in order to redirect heat away from the interior, reducing the need for air-conditioning. Made up of structural insulated panels (SIPS), the Element House is designed according to Passivhaus standards and operates independently of public utilities by integrating passive systems and on-site energy generation.

Fig. 2.55 Element House at night. *Architects* MOS Architects. *Photographer* Florian Holzherr

Fig. 2.56 Interior of Element House showing kitchen

Fig. 2.57 Plan of Element House

2.6.2 *Casa GG Montseny—Barcelona, Spain, 2013. Alventosa Morell Architect*

Timber frame

The client requested that the design should be economical, built in a short period of time and be energy efficient as a Passivhaus. The client also stipulated that there should be six cells providing three bedrooms, a kitchen/bathroom cell, a living space and a store for bikes, etc. In between these cells, there should be room for dining and study with scattered provision of shelving for their large book collection (Figs. 2.58, 2.59 and 2.60).

These cells have been disseminated from the traditional compact house form that we are all familiar with. The interconnecting spaces are useful buffer zones between each cellular activity. As a result of the distancing between the cells creating interesting recessed modelling around the perimeter, the plot size is larger than the equivalent compact form of building.

TECHNICAL

Timber was used throughout as an obvious organic connection with the surrounding woodland, as well as being economic means of construction. The house uses 76.77% less energy to heat the house, in comparison with the same house built in the traditional way.

Fig. 2.58 Casa GG Montseny. *Architects* Alventosa Morell. 2013. *Photographer* 69Goula. *Courtesy* The architects

Fig. 2.59 Interior showing book shelving integrated into the wall

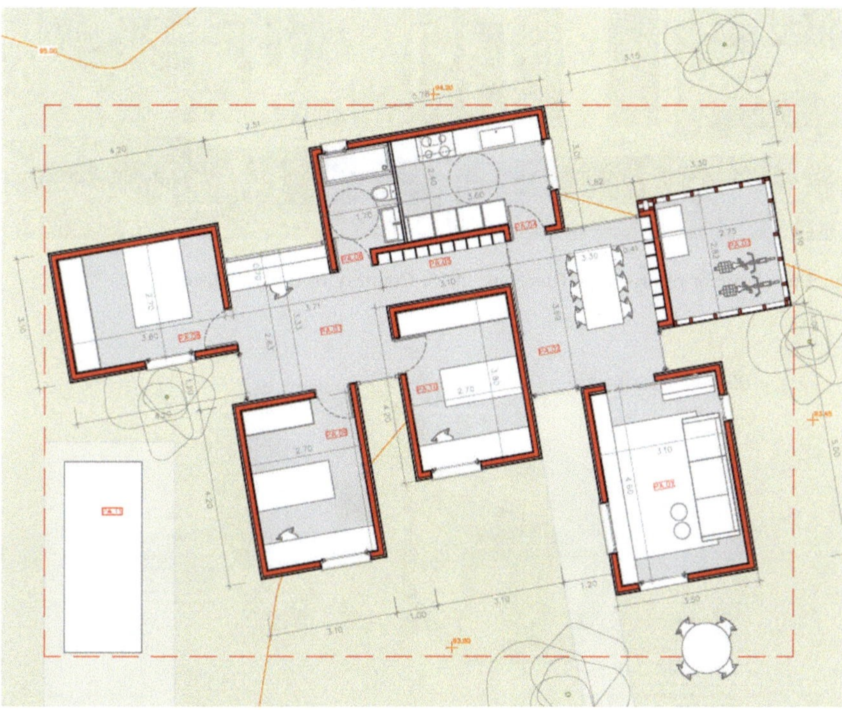

Fig. 2.60 Plan of house

2.6.3 *Hivehaus—Prefabricated Modular Units*

Here is my first example of a modern cellular design based upon the hexagon. Designed by Barry Jackson from Wigan in Greater Manchester. He was inspired by its organic existence in the beehive, hence the name 'Hivehaus' with the 'Haus' taken from the German word for house as a homage to the Bauhaus. He also wanted to build some homes that would be far more affordable than traditional brick and mortar types of house (Figs. 2.61 and 2.62).

Fig. 2.61 Exterior of prototype Hivehaus. 2015. Courtesy of Barry Jackson

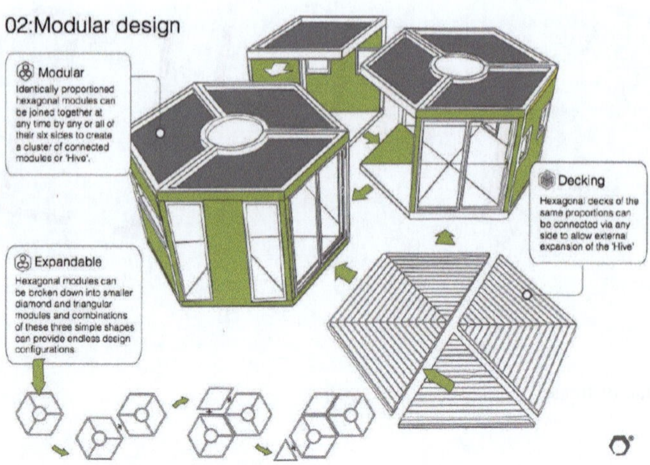

Fig. 2.62 Fact sheet taken from website

Fig. 2.63 Interior view across two cells

The system does not include any built-in furniture as that is left to the client to furnish. Each cell amounts to about 10 m^2 and each side is 2 m wide. It is made of standardised components that enable the cells to be connected to each other to suit a variety of layouts which can incorporate work/live space, e.g., office module, lounge module, kitchen module, or sleep module. The components are broken down into equilateral triangles and permutations of these. The wooden frame floor sits on feet that adjustable to compensate for uneven ground. Electrical wiring runs through the floor and lights are remote controlled so there are no switches in the walls (Fig. 2.63).

2.6.4 Earthbag Building Co, USA

This is a true Eco Home by using locally sourced earth and filling bags with it. Earthbag building fills a unique niche in the quest for sustainable architecture. These bags lower the embodied energy commonly associated with the manufacture and transportation of building materials. The fill material is generally of mineral composition and is not subject to decomposition (even when damp), attractive to vermin, or burnable…in other word it is extremely durable. The fill material is generally completely non-toxic and will not offgas noxious fumes into the building. The structure is typically finished with plaster internally, and stucco externally to shed water and to prevent any degradation from solar radiation (Fig. 2.64).

Because the earthbags can be stacked in a wide variety of shapes, including domes, they have the potential to virtually eliminate the need for common tensile materials in the structure, especially the wood and steel often used for roofs. This not only saves more energy (and pollution), but also helps save our forests, which are increasingly necessary for sequestering carbon (Fig. 2.65).

The above plan of three hexagonal pods and two interlinking squares shows three bedrooms, one with a walk-in closet. Remaining spaces are a plant room, two bathrooms, a laundry, an open-plan living, dining and kitchen space. There are also two bonus rooms without doors, which are offered up for use as a study or any activity that suits the occupants. The plan respects the hexagon as a controlling geometric element, but it also provides 90° walls to form rectangular spaces. The whole idea depends upon a sympathetic landscape that can provide these materials.

Fig. 2.64 Earthbag. A visual of the exterior

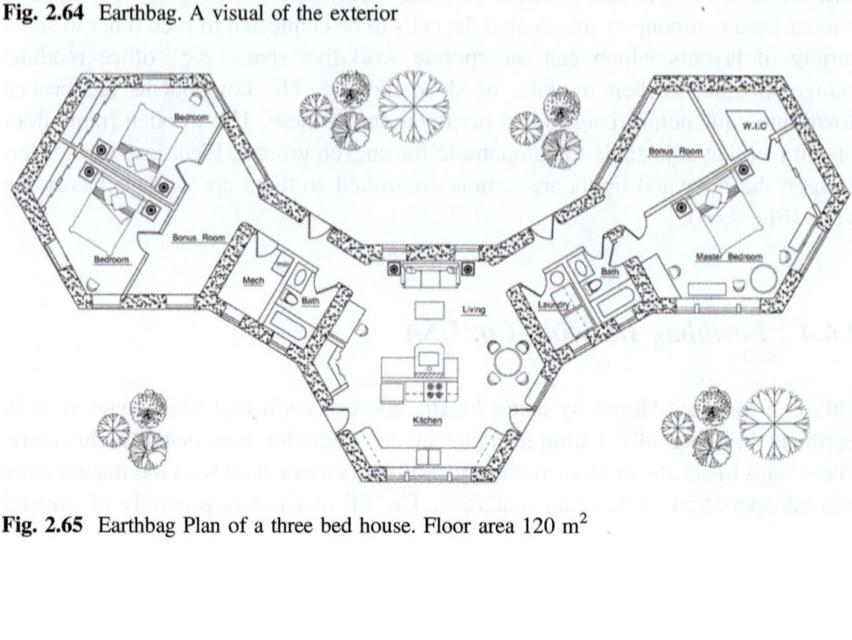

Fig. 2.65 Earthbag Plan of a three bed house. Floor area 120 m^2

2.6.5 *MM House, Palma, Balearic Islands, Spain. Architects OHLAB*

This project, built to Passivhaus standards, groups four boxes—kitchen, living-dining, main bedroom and guest bedrooms—which can be used together or independently. Each box is placed carefully on the ground and rotates on its axis to find the best views and orientation for their use. The bedrooms face the East, the garden and the Bellver Castle; the living and dining room face south-east, the sea and the garden; the kitchen faces South and the vegetable garden; and in the attic over the living room, the terrace looks towards the view of the sea. Floor area is 196 m^2 (Figs. 2.66 and 2.67).

Fig. 2.66 MM House, Mallorca, Spain. *Architects* OHLAB. All photos by José Hevia. *Courtesy* The architects

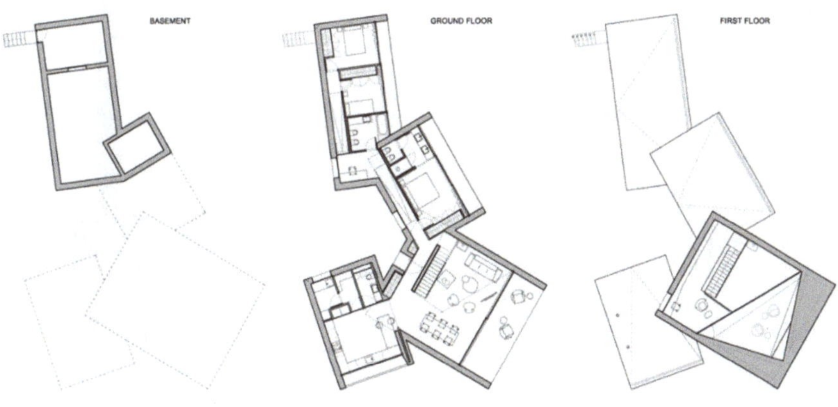

Fig. 2.67 Plans of MM House

The entrance to the house is almost concealed. It comprises two double bedrooms and one single bedroom. The apparent haphazard planning reveals a careful control over an almost semi-circular circulation route from the bedroom wing via the hall and living space and finally the kitchen. There is one en suite bathroom, one family bathroom and one WC off the kitchen adjacent to the utility room. The over spacious living/dining area has an exposed staircase going up to a study area and roof terrace (Fig. 2.68).

TECHNICAL

A rigorous infographic and thermal study has been made to ensure an optimal heat input, maximising it in winter and minimising it in summer. The façade has an exterior insulation system that increases the insulation thickness up to 15 cm and covering all joints to completely avoid any thermal bridge. The heat exchanger ensures air replacement wastes no energy, and it is optimised to use the heat generated from showers' steam and cooking. Thanks to this design cooling is not necessary, which is typically an issue in the hot days of Mallorca's summer, and the heating demand is only 11 kWh/m^{2}a (a typical house of this characteristics in Mallorca would have a heating demand of 85–100 kWh/m^{2}a). This is partly due to a perimeter underfloor heating circuit. The energy to heat the hot water is obtained almost entirely from the solar thermal installation.

The pitched roofs have a system for collecting rainwater, three of them get water for irrigation and general use whilst the fourth roof is 'the clean one' and collects

Fig. 2.68 MM living/dining area

water for consumption. For this purpose, there are two separate tanks, one of 40 m³ and another 8 m³, which are arranged taking advantage of the gaps created between the housing and the slope of the terrain. The garden includes a vegetable patch, with low-maintenance native vegetation and deciduous trees along the south of the house to protect from the summer sun.

2.7 Adjustability

When examining the requirements of the home it is important to realise that with the passing of time the needs of a family will change which will mean that the existing furniture, resources and space use will also change. This suggests that some degree of adjustability or flexibility of facilities should be built into the design concept of a house. If such adjustability is not possible, this produces frustration on the part of the occupants and the desire to move is increased. We have witnessed this many times in the UK. Here are some of my own observations that need attention.

A. When couples first set up home, the concept of 'sharing' becomes a difficult hurdle to get over. They begin to sort out the three domains for the sake of staying peacefully together: His, Hers and Ours but space use needs working out.
B. It amazes me that when babies arrive their crawling habits seem to come as a surprise to the parents, and they get upset when the baby goes to places that it should not go to. There is no adjustability with the furniture to make allowances for this. A playpen is the usual answer.
C. The hinged door needs serious review and is why it is one of my four no's. This is a dangerous heavy 2 m × 0.8 m panel which causes children to trap fingers, or it is used to slam when rage occurs. When it is closed, it forms part of the wall, but when left open, it becomes an unwanted intrusive element in the room. It is a rotational device but the act of rotation has no design expression whatsoever.
D. Repeated usage of furniture and surfaces makes them worn. If this is undesirable then instead of replacing the item why not rotate or reposition to allow for the remainder of the item to receive equal wear and tear? Recycling in one's own home is to be encouraged for disposal is too easy, adds to waste and incurs more expense on replacement.
E. Change of use examples: Garage becomes a workshop/store. Ground floor toilet becomes overloaded with cloaks, wellies, and anything else because the hall is too small. A child's bedroom slowly changes into a study/living space as the main activity, but initially it is only designed and named as a bedroom. Lounge suffers from reduced usage because the dining/kitchen has become the central hub of activity and communication.

The following projects, which are a mix of a house and apartments, show examples of adjustment and flexibility to suit the requirements of the occupants.

2.7.1 *All I Own House, Madrid 2014. PKMN Architects in Madrid. All Photos Javier de Paz García*

The owner inherited an existing single-storey building with large garden from her grandmother. She decided to convert it into a single-bedroomed dwelling in such a way that it would allow flexibility of use for her and her partner which is required on a regular basis (Figs. 2.69 and 2.70).

The whole floor area was divided into two halves: one half used as a multi-purpose open space and the other half to cater for the kitchen, dining, sleeping, study, bath, and toilet. This was achieved by the design of three chipboard made shelving storage units that slide on industrial glides fixed to the top of the wall on one side and to the downstand beam on the other (Figs. 2.71 and 2.72).

Fig. 2.69 Madrid house view of multi-purpose space on left

Fig. 2.70 View showing how main storage units are adjusted

Fig. 2.71 View showing the bed hinged down

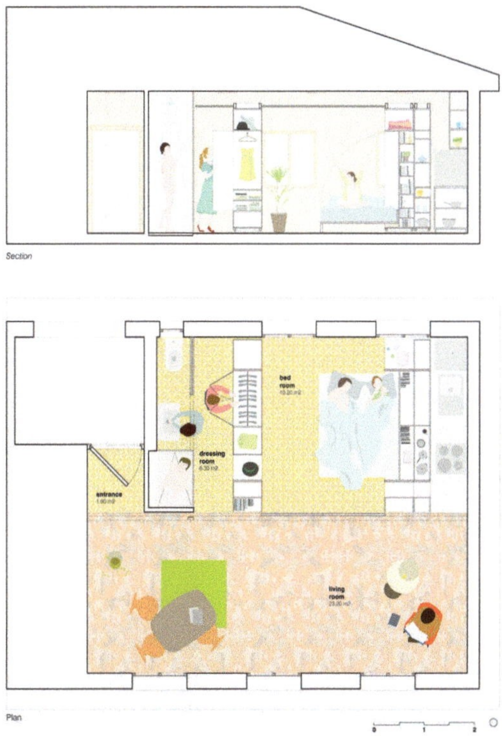

Fig. 2.72 Plan and section

My main criticism is that the whole setting is in a constant state of unrest due to the impending move of the units, which is clearly visible.

2.7.2 *Susaloon—Refurbishment of Apartment in Madrid, 2014*

Architects: elii—Uriel Fogué + Eva Gil + Carlos Palacios.

The aim here was to design spaces that could be changed and adapted for a variety of uses.

This was achieved by the integration of a series of fold-down and sliding devices using timber framing that is used to divide the spaces, deploy a work space, a large dining table, a bed for a surprise guest, a spacious area to practice shiatsu or to reorganise the storage space. The dominance of the timber grid is rather more successful as an integrated interior than the Madrid house (Figs. 2.73, 2.74 and 2.75).

Fig. 2.73 Framed cladding in closed position

Fig. 2.74 Cladding in open position

Fig. 2.75 Another space showing components lowered into position. *Photos* Miguel de Guzmán and Rocío Romero

2.7.3 University of Buffalo's Grow Home—2015

An Eco-Friendly Prototype whose Furniture Maximises Energy Efficiency.

The US Department of Energy's annual Solar Decathlon called upon collegiate teams to design and build cost-effective, energy-efficient homes blending consumer appeal and optimal energy production. University at Buffalo, The State University of New York, rose to the challenge with their GRoW Home—a 1100 ft^2 of space focusing on self-sufficiency and versatility. Within UB's large Solar Decathlon team of more than 200, 17 students led by assistant professor of architecture Nicholas Rajkovich sought to really push the envelope (literally) with custom furniture.

> *The GRoW Home seeks to better connect the resident with the natural environment. The building envelope employs structural insulated panels, two high-performance sliding glass doors, and a series of fixed superwindows to reduce heating loads. This 8 ft tall shelf partitions the bedroom and dining area, and features a rotating TV screen allowing both rooms viewing access.* Assistant Professor Martha Bohm, Dept of Architecture.

Figures 2.76 and 2.77 show the installation of these sliding partitions.

Fig. 2.76 GrowH sliding wall panels in open position

Fig. 2.77 GrowH sliding wall panels in closed position

2.7.4 Apartment in Brera, Milan, Italy. 2015. Designed by Planair

This tiny apartment (34 m^2) is making the most use of a small area by concealment of storage and a bed whilst achieving privacy when necessary. Flexibility is promoted by means of different rooms through moving partitions which can be closed or opened thus producing different spatial combinations. The furniture contains utensils and functional elements which allow the spaces to be inhabited according to requirements (Fig. 2.78).

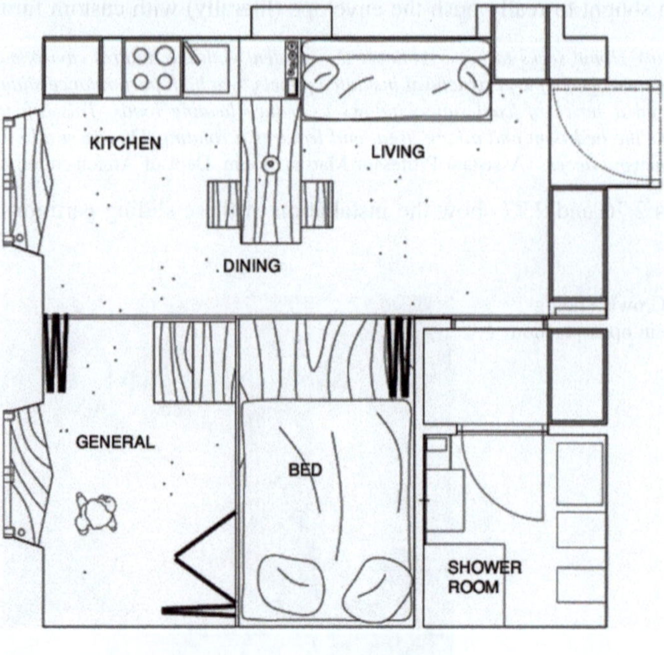

Fig. 2.78 Plan of apartment in Brera

Fig. 2.79 Four photos of apartment showing how adjustment is achieved

There is a chromatic range of wall colours that follow and strengthen the qualitative differences of different private spaces, starting from bright and white ones in more public areas and continuing with a more saturated and dense grey in more private environments. Natural ash wood is used throughout (Fig. 2.79).

2.7.5 Ori Systems, Cambridge, MA, USA, 2016

Ori's residential systems offer intelligent space on demand: with the light touch of a finger, a living room converts into a bedroom, a walk-in-closet, or an office, all connected to any smart-home ecosystem. Ori's innovations build on research at MIT's Media Lab focused on using technology to respond to the challenges of global urbanisation, led by MIT Professor Kent Larson and his CityHome project Media Lab founder and technology visionary Nicholas Negroponte who was actively involved in the conception of this research project and is currently an advisor for Ori.

> *The inaugural Ori system is a robotic furniture system that transforms at the touch of a button. Tucked into the bottom of the unit is a comfortable bed* (see Fig. 2.80) *that glides in and out. When tucked in, the bed is hidden under a closet, office and couch to maximize space. When activated, the whole unit (b) slides in and out to become a bedroom or a more generous living room. One side of the unit hosts a full closet, but also hides a desk (c) for a home office; the other side holds a media center for entertaining. Each room can be preset for a space, including preferred integrated lighting, so that one touch on the physical*

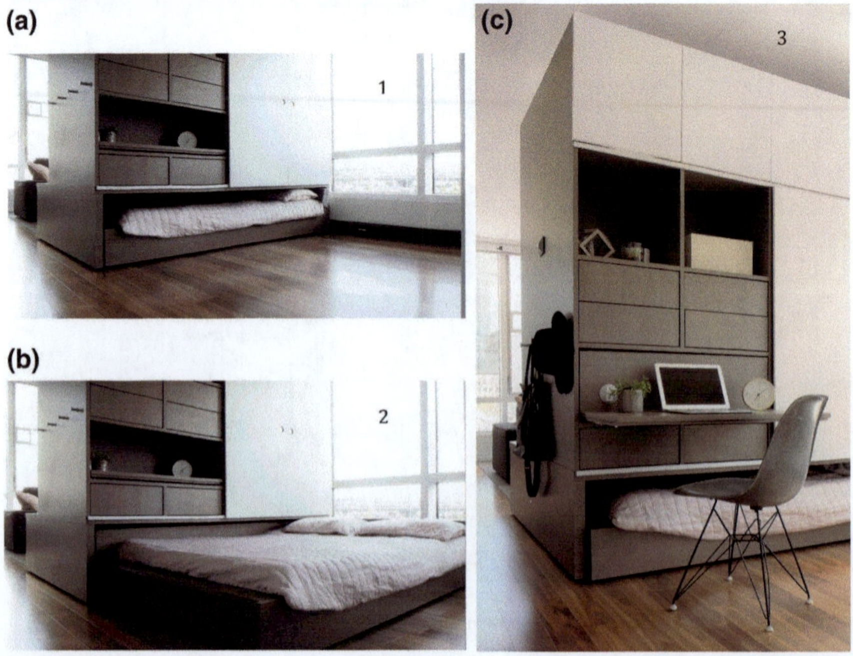

Fig. 2.80 Two photos **a** and **b** showing a bed revealed, and **c** showing pull out desk surface

interface or on the corresponding app will morph the room. The on-device console uses manual glider buttons that make the owner feel as though they are magically moving the unit. Yves Béhar

Whilst the explanation of this project is appealing, the visual result appears to be rather more heavy and cumbersome even though it is aimed at the luxury end of the market. The bed is very low down which is not ideal for people with limited mobility.

2.7.6 Zoku Hotel Concept, Amsterdam, 2015

This is outside my category of housing, but I have selected it for its ingenious use of space. Hans Meyer is co-founder and managing director of Zoku, a re-invented apartment hotel concept, facilitating global living and working for the independent travelling professional. At 46 m^2/495 ft^2, the Zoku Loft is the largest Loft, which features an adjacent extra bedroom. The second sleeping space has extra wide doors and a wheelchair accessible bathroom and shower area. A large kitchen table serves as focal point with Muuto design furniture elsewhere. The elevated sleeping area can be screened off for visitors. Sliding doors can also hide away the bathroom's shower. There is plenty of storage (Fig. 2.81).

Fig. 2.81 Photo on left shows the stairs in closed position. Photo on right with stairs open for use including door access to bedroom. Courtesy of Zoku. Photos copyright of Ewout Huibers for Zoku and concrete

2.8 Furniture

The previous examples in Sect. 2.7 were designed as interior fittings each belonging to a dwelling. The following examples are products un-associated with any particular dwelling, and because they have some degree of flexibility about them, they may help in my examination of house interiors.

2.8.1 LIND from DIZZCONCEPT by Designer Darko Špiljarić, Zagreb, Croatia

Lind is a multi-functional modular system of storage/display elements of perforated MDF panels for wall cladding and partitioning. These range from boxes, shelves, drawers, tables, to hangers capable of being adjusted with out using tools (Fig. 2.82).

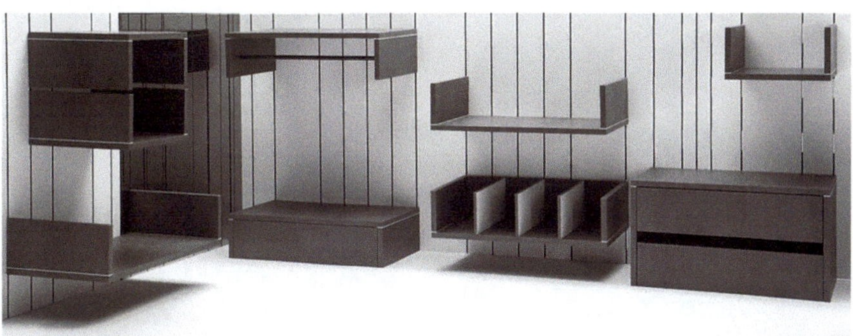

Fig. 2.82 Lind range of modular adjustable wall furniture. *Courtesy* The designer

2.8.2 *Folding Wall Hung Table 2015*

Bogdan Ciocodeică & Diana Roşu two architects based in Bucharest, Romania together with architect Attila KIM created this innovative poplar dining table that can be folded into the wall when not in use. Perfect for small spaces, the table is made from a single piece of wood that can be mounted in the wall and the table 'fold out' from it. The beauty of this is the 'footprint' left on the wall when the table is folded down. It is not just a folding table but it provides a wall feature as well, making it have two lives (Figs. 2.83 and 2.84).

Fig. 2.83 Table's profile clearly seen on the wall. *Courtesy* The architects

Fig. 2.84 Table folded down for use with hinged legs in position

2.8.3 *ARRAY Designed by Tijs Gilde of the Netherlands. 2016*

ARRAY is an exploration on creating diversity in the living room setting, Tijs Gilde sets out to break the conventions surrounding the composition of todays living room. Instead of having the usual tv-carpet-couch setting he suggests a more flexible alternative. Soft shapes made from foam are upholstered with high-quality felt fabrics giving a sense of luxury.

Each block is provided with a yellow strap that pierces the objects, allowing the user to strap different blocks together into new compositions. It is a shame that there are no figures using them to give a sense of scale. Whilst they are delightfully fresh in concept, my suspicion is that they will soon be dominated by wild tendencies to throw them around creating chaos. But this is maybe the desired result (Fig. 2.85).

Fig. 2.85 Array foam furniture in a variety of groups. *Courtesy* Tijs Gilde

2.8.4 BOXETTI Designed by Roland Landsberg from Latvia, 2008

This is a multi-functional kitchen island unit with a sink and an automatic sliding surface to reveal oak cutting board with built-in spotlight above. Tableware and kitchen knives can be stored. It is equipped with two extendable bar seats a built-in fridge, shelves and drawers. All painted wood (Figs. 2.86, 2.87, 2.88 and 2.89).

The attraction of Boxetti is the compactness of all the various components in carrying out so many tasks. This follows with the seductive angularity of the units, which are screaming to be let out of the stranglehold of the box (Echoes a theme

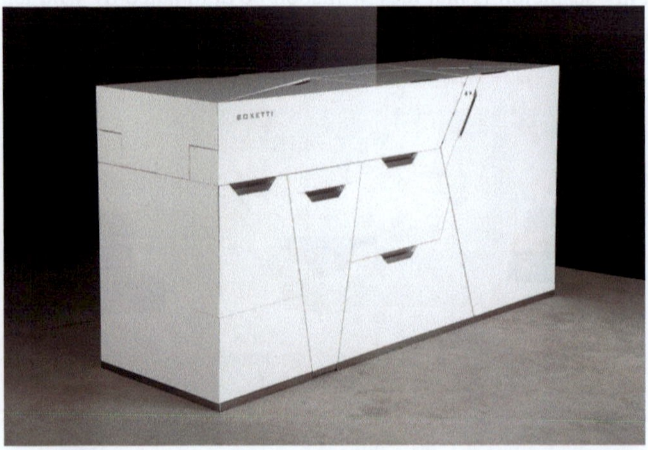

Fig. 2.86 Boxetti main closed box

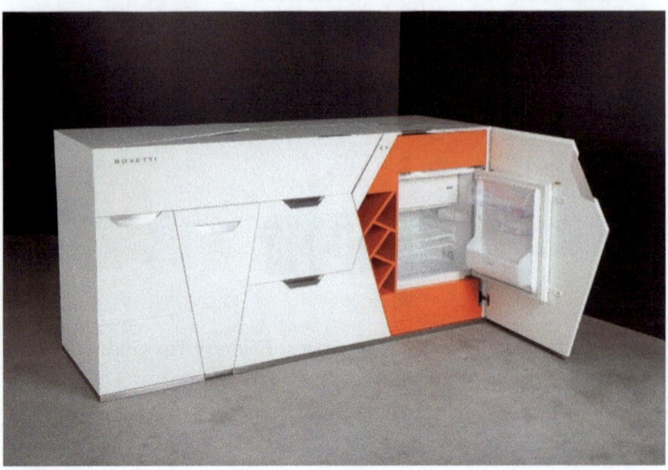

Fig. 2.87 Revealing fridge

Fig. 2.88 Revealing chopping board and seat

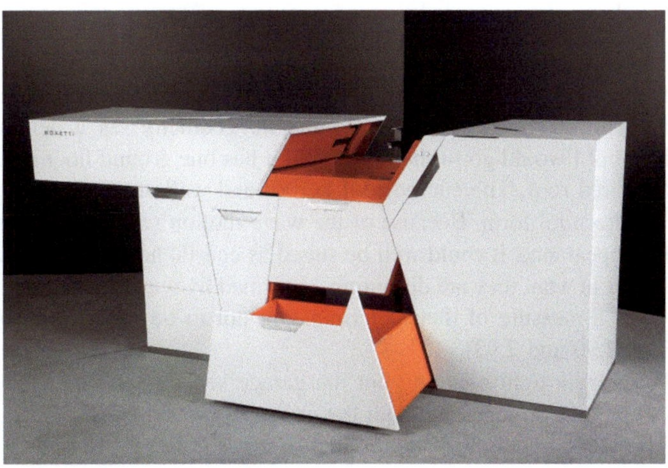

Fig. 2.89 Revealing sink

here from somewhere!). The ultimate test and probably sacrifice is when performing the various functions the question has to be asked: is this the best way to perform this function? If the answer is an annoying no, then I am afraid the design fails and will be cast into the 'curious oddity' category—for amusement only.

2.9 Future Project

This is a project that is awaiting a commission or seeking planning development.

2.9.1 The Arc House

Detached three bed house by Architects McInness Gardner of Glasgow.
Winner of the Sunday Times Homes for Britain Competition 2015.

Architects' statement:

- *Our aim was to produce design a that created a pleasing and inviting geometry and to include enough detail to engender a sense of intimacy, security and belonging.*
- *The front of the house to be minimal in size to provide the practical and social interface with the community, allowing a larger more private and green relaxing garden to the rear with connections to the cycling and pedestrian network.*
- *The house we have designed comprises 3 floors and maximises use of the whole structural volume of the house and incorporates the latest smart green and sustainable technologies. Our vision was to create a more romantic geometry and roofscape within a treed setting.*

The most stunning aspect of this design is the external form reminiscent of Dutch gable ends or primitive huts (see Figs. 2.91 and 2.92), which defies the economic logic of stacking one floor on top of another even though the floor layout is different. The architects refer to a 'romantic geometry' as a defining visual characteristic but I would go further by saying it has one ground floor external wall and a two-floored roof. The curved roof is roof and wall at the same time—not a common architectural form. Because of the wave motion seen down the street and its boat hull appearance, it could well be suited as coastal homes. For the architects to have a belief in what they are designing confirmed by winning the Sunday Times competition is a measure of their determination not to be sucked into box architecture (Figs. 2.90 and 2.93).

The plans reveal tight planning but the garage sensibly accommodates utility, power unit and bike storage, although the utility space and comfort level is small. Storage is provided under the stairs (dogleg so that's good) for coats and shoes but

Fig. 2.90 Drawing of street view of the houses. *Courtesy* McInness Gardner

Fig. 2.91 3-m-high Hut of Toda tribe in India. They are built of bamboo fastened with rattan and are thatched. *Wikipedia* Pratheep P. S.

it is still a little cramped. Bedrooms have excellent amount of space, and there is a study on the top floor, although as I have said before workspaces need to be on the ground floor as part of the general hub of our shared digital communication era. The upper floors will feel the impact of the sloping walls but not to any great detriment. The main disappointment is to follow the trend of lumping the kitchen, dining and lounge into one space. It may work in a large open-planned space where people are not on top of each other in their separate activities, but in this confined space irritation will be felt by the occupants. The designers have shied away from laying out a dining table and chairs plus more lounge seating for watching TV for example. Lastly, the lack of general storage is my main criticism although I can see the garage taking some of the share of that facility.

Fig. 2.92 American Dutch barn. *Courtesy* Kate Kelly. *Photographer* Howard P. Mock

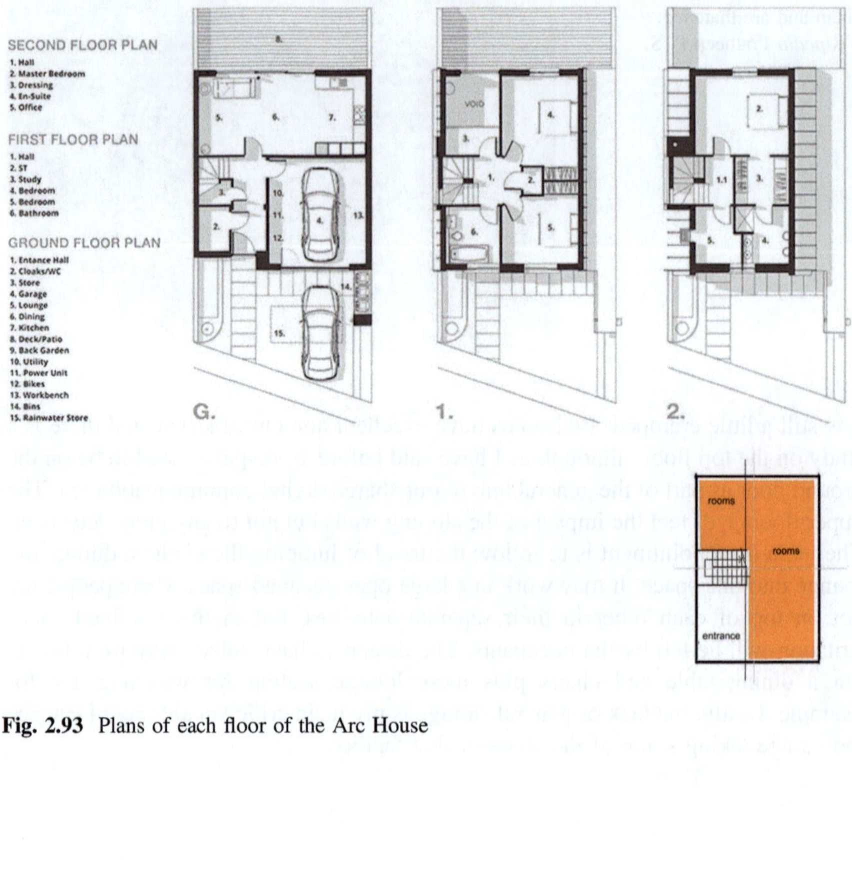

Fig. 2.93 Plans of each floor of the Arc House

2.10 Cala Domus Housing, Newhall, Harlow, Essex, UK, 2008

Architects PCKO.

2.10.1 Description

I must confess that I had never heard of this popular development of ten years ago until researching for this book, and it has taken me by surprise. Mainly because of the notion of having architects selected to design different parts whilst following the master plan of Studio REAL. This is in contrast to many developments, which are usually in the monotonous style of one builder. Therefore, we have a neighbourhood of well designed contextually varied houses and apartments with well-planned access roads and special parking spaces adjacent to all dwellings. I like this particular scheme Cala Domus because it is so refreshingly different to the mass of housebuilding being done at the present time. I visited Newhall and was blown away with its design ingenuity. I spoke to one resident of this house type and he had lived there since it was built and was perfectly happy with the house, apart from one criticism: due to amount of glazing to the front elevation he experienced so much heat loss that it could be colder in winter.

Cala Domus is an award-winning residential scheme for Cala Homes. The development of 74 homes is laid out at a density of 39 dwellings per hectare and consists of houses, flats, workspaces and duplex apartments. Laid out in pedestrian-friendly streets and lanes, all homes are within 400 m of a bus route and within 60 m of parkland. PCKO designed an innovative wall system that incorporates a one-metre strip to house services and provides storage for refuse and recycling.

The homes are distinctive and colourful, combining glazed balconies, coloured render and traditional brick. The construction method used is a highly insulated off-site manufactured timber frame and roof panel envelope (Fig. 2.94).

The plans below reveal a well thought out layout though I feel that the dining activity would be a better partner to the kitchen, as it is too far away from it. The bedrooms are a decent size with good distance separating each one giving great privacy, as opposed to so many houses built with bedrooms separated by thin walls. The study at the top of the house has plenty of daylight and is surely big enough for more than one person and a delightful space it is. I noticed that for some houses it appears to be a multi-functional space with gym equipment for example. The garage is more like a car port being open to the air at the rear and a door/gate to the front. When I saw the name I thought that it must have some strong connections with Majorca! (Fig. 2.95).

Fig. 2.94 View of the front of detached houses of Cala Domus. *Photo* Author

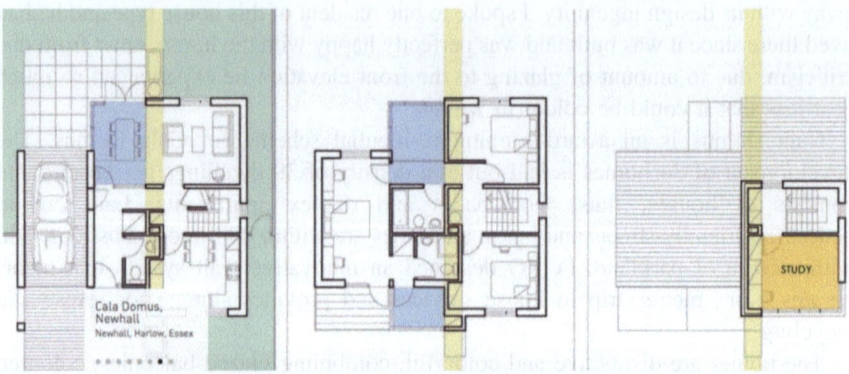

Fig. 2.95 Ground, first- and second-floor plans of Cala Domus

2.11 Conclusion

It is quite apparent when comparing all of these projects together that a different set of priorities exist for each one, which means that they are not all in the same race. But they are providing some form of housing and furniture for people and those occupants will indeed vary in terms of socio-background and financial status. Some projects will cost more than others in terms of cost per sq metre but costing becomes quite a difficult comparative exercise when taking into account land price variations, and local labour costs. I have included a section on furniture because it is inextricably linked to how a space is used, whether it is built-in or freestanding. Therefore, what can we learn from these Case Studies?

I have seen some ingenious planning solutions handling very tight spaces. There are some well designed and innovative building forms using a variety of materials.

The Newhall project stands head and shoulders above anything else I have seen. The technology of building and services will have to be measured in the light of the contents of Chap. 6. The major disappointment for me, and this substantiates why I have written this book, is that the interior spaces provided for general mass housing, including the award winners in this chapter, is still too limiting when it comes to maintaining a sense of human dignity. The trend of open plan containing the kitchen, dining and living areas smacks more of economic necessity than honouring the needs of the occupants. Feeling cramped is one of the worst living conditions human beings have to endure and it can affect mental as well as physical health (see Sect. 4.5.4). A lack of privacy and storage are clearly evident and in order to redress these there must be an increase in the amount of space provided. But that is too simplistic an answer to the problem. Many of the 'neighbourhood' schemes show how the architects have used their ingenuity with the external modelling of the houses offering a variety of juxtapositions and visual treatments mixed with some delightful landscaping. But the interiors are still the same boxy spaces devoid of any design innovation that the 'architecture' has been given. The left-hand diagrammatic elevation of terraced housing in Fig. 2.96 has become sadly far too common in many housing schemes being reminiscent of the severity of Nazi architecture or prisons.

If we compare the main inventive changes made in housing with that of the motor car it is quite clear that whilst the car has been finely tuned since first invented in the late nineteenth century and open seating to provide us with abundant choice, fast speeds, great comfort and indoor entertainment, the housing market is so dull by comparison. Since the houses built from the 1900s, we still commonly use such core terms as kitchen, bathroom, hall, utility, living room, dining room, study and bedroom, whilst other terms such as scullery and parlour have been superceded. Wealthier homes would have a music room, a games room, a bar, a conservatory, (curiously, whilst this has become a very common extension to many houses, developers do not generally supply houses with these) a gym and a library (closest thing to a study) reflecting their need for such spaces. The sad thing is that more humble families also may need such spaces and have to make do by using the garage or a spare bedroom if there is one. Of course, we have made technological inventions with regard to building services, products and construction, but we have not responded to the changes in modern living patterns by updating the interior space provision (see Sect. 3.2.3 for new names).

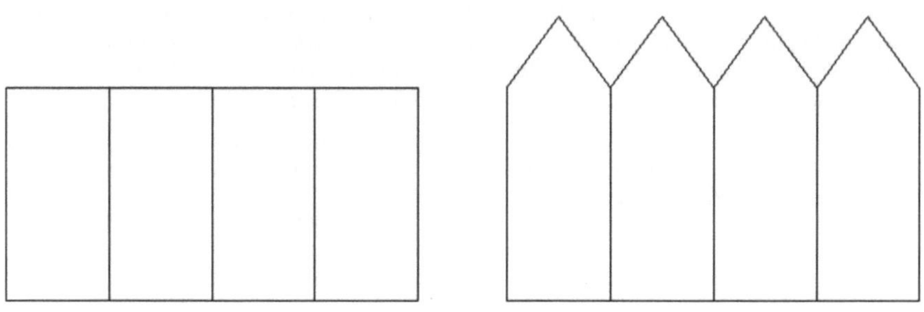

Fig. 2.96 Diagrammatic basis for typical modern terraced housing elevations: left lacks definition, right is preferable

I have seen many new estate houses and some are shown in this chapter. I have been appalled at the poor standard of construction of some of these dwellings especially with regard to the provision of more modern facilities and equipment. As Alex Proud observes:

> *But these dreadful developments are repeated on the outskirts of every town from Land's End to John O'Groats. Mean little brick boxes dotted around dead end streets with no shops or amenities. It's not just the layout either, it's the quality. Outside, you'll have crap bricks, crap concrete tiles, crap plastic windows and crap detailing. Enter through the PVC door with its twee Edwardian glass, noting the plastic carriage lamp. The hallways will have cottage height ceilings, with a faux Victorian staircase and stunted Georgian skirting boards made of MDF. The living room will boast a fire surround that apes some period between 1720 and 1920 and looks like it was made on a 3D printer. And so on.* Alex Proud[5]

The scandalous and undignified conditions that some people live in, whether in rented or owner occupied properties needs serious attention. There are bodies already involved with this sector known as 'affordable or social housing' whereas I am dealing with the mass estate housing sector serving the majority of the population. The charity Shelter has this to say about the conditions of the existing housing stock which new housing is supposed to solve:

> *More than four in 10 homes in Britain are falling short of an acceptable standard to live in, according to an alarming new report from the housing charity Shelter. Nearly one in five homes failed the Living Home Standard based on a lack of decent conditions. Some didn't have running hot and cold water, others were not structurally sound, and many had serious pest infestations or issues with mould and damp.* Shelter[6] Independent, 17.10.16

When I speak to builders and developers about this state of affairs, they point to the current legislation over the value of land. Unfortunately, land price increases affect the market value of a house that a builder can sell it for in order to make a profit. Builders will agree that the price of building a three-bedroomed house will be the same wherever it is built, allowing for variations in local trade prices. But the selling price of that same house will fluctuate according to the desirability/wealth of the neighbourhood. Regarding the construction quality, this is down to meeting the budget to build. The Guardian comments again:

> *They might be the most ubiquitous feature of the modern English landscape, and yet they barely attract any comment: those sprawling newbuild housing developments that seem to surround almost every town and city, offering a promise of comfort and security and a vital foot on the property ladder. This week the Guardian reported that Bovis is set to award people who live in some of its newbuild homes a total of £7 m in compensation, in response to claims that houses have faulty plumbing or wiring, missing insulation, and other serious defects. Some people say they were offered money to move into homes that have not been completed. When the news broke, the Bovis share price fell by 10%, wiping £100 m off its stock market value.*
>
> *This is just one part of a bigger story of complaints about Britain's construction giants – and what happens when the rush to build leads to corners being cut and houses left either unfinished or deeply defective.*
>
> The Guardian 24 February 2017

[5]Alex Proud, *Why is Britain so terrible at domestic architecture?* The Telegraph, Sept 2014.
[6]Shelter—The UK Housing and Homelessness Charity.

One of the disappointing results of my research, despite some schemes receiving housing awards, is the almost formulaic repeat of the same or similar plan and a lack of imagination with the interior design of the dwellings. There is also the unexplained habit of increasing the living/dining spaces in proportion to the number of bedrooms in the house. As will be seen in my own design proposal in Chap. 5, Sect. 5.4, the ground floor spaces remain the same area because no matter how what the size of family, certain facilities and space will always be required. It is back to the 'status' thing of proportional equality throughout. Even with interesting mixed developments and landscaped walkways the end result is very boxy and uninspiring. Clare Nash puts it well:

> *Our creative talents are highly sought after in fashion, music and architecture. We embrace technological change at alarming rates. However, I have been getting more and more frustrated about how little we apply these talents to our housing. Our town edges have become submerged in a sea of red brick, on forms that looks the same whether you are in Southampton or Northampton. As British people, we need only drive an hour or so away to hear a very different regional accent. And yet regional variation in new build housing has all but disappeared. The UK is a rich composition of diversity, why can't this be evident in our housing? It doesn't seem fair that the majority are required to live in identikit brick boxes. These homes have no sense of place, do not encourage community and offer extremely low standards of design and space. Residents have told me they feel 'contained'.*
> Contemporary Vernacular Design. By Clare Nash, Published RIBA Publishing 2016.

Nash abhors the lack of regional variation in the design of estate houses, which is true, but there is an attempt by some architects to apply some individuality in their designs as can be seen by these Case Studies. In fact whenever the local vernacular influences the design of housing it can sometimes result in a pastiche of the worst kind such as is shown in Fig. 2.97. Although this development is worthy for providing affordable housing that the Sanctuary Group is well known for, it is a pity that it is

Fig. 2.97 Wolfson House, Pimlico. 24 new build mock Georgian terraced homes by Sanctuary Group, 2014. Photo courtesy of Inside Housing

mock Georgian rather than being a modern building. The curious thing is that the original buildings that were there are associated with the wealthy beyond the reach of ordinary folk, yet this development is for the lower income group of the community.

Summary of Comparable Floor Areas

Interwar semi	67 m^2	Three bed
Prefab	59 m^2	Two/three bed
Harlow end terrace	72 m^2	Three bed
1950s detached	191 m^2	Five bed
Span terrace	84 m^2	Four bed
New Ash Green	130 m^2	Five bed
Tudor Walters Committee	98 m^2	Three bed
Oxley Woods	81 m^2	Three bed
House	93 m^2	Three bed
Barratt	74 m^2	Four bed
Villa Verde	85 m^2	Four bed
Abode, Kneighton	173 m^2	Four bed Less 25 m^2 for terraces = 148 m^2
Uji House	103 m^2	Three bed
Balsthal	192 m^2	Four bed
Hanse Haus	170 m^2	Three bed
Karst House	85 m^2	Two bed
Zedfactory	139 m^2	Four bed
Orchard Close	234 m^2	Four bed
Letchworth	105 m^2	Three bed
Isle of Wight	109/114 m^2	Four bed
Murphy House	165 m^2	Three bed
Mami House	162 m^2	Three bed
MM House	196 m^2	Three bed
Earthbag	120 m^2	Three bed

Chapter 3
Living in the Home

Abstract I should explain that this approach and the subsequent design solutions I am proposing I hope will be a contribution to the problem of estate house design. I have obtained feedback along the way from a variety of professionals, and whilst many have challenged my ideas, in general they are supportive. Before any design proposals are considered, it is important to understand the functions of the home and how they have changed over the past few decades. The domestic activities are examined in two ways: firstly, in relation to each other spatially, and secondly, the kind of facilities necessary to support those activities. Out of all the sectors in the design industry, domestic design is probably the most problematic in the sense that everyone lives in a home and it is so easy for designers to make assumptions based upon their own experiences. The failing of 'design' today in a general sense is well put by Alison J. Clarke in the Introduction of her book 'Design Anthropology': *Observational techniques, human focus and emphasis on the machinations of the every day are essential in interpreting the complex implications of consumer culture, technological interaction, and media. As the values of expanding new markets challenge homogenous, globalised understandings of product worlds and users, the desire for indigenous, grassroots, and nuanced insights has never been more acute.* This of course can dangerously make a design proposal prejudiced without the proper unbiased research to support it. I could be guilty of being in such a position, but I would quickly come to my defence by saying that a lifetime's work of study, designing, teaching, observation and a missionary zeal to make things right should place me in a good position. I am well aware that some of my recommendations have never been made before (to my knowledge) or that they may have a mixed reception, as long as something constructive comes out of the book. In designing an estate home for the masses, there is no Mr and Mrs Average, and there is no such thing as a tailored solution except if you were designing bespoke for one client. For example, family life is complex and varied depending upon income group, culture, religion and working role in society. Therefore, decisions and conclusions are drawn up that would appear to cater for all, or at least provide a basis for adjustment to suit.

A. Sully, *The Estate House Re-designed*,
https://doi.org/10.1007/978-3-319-90397-2_3

3.1 Major Considerations

3.1.1 Forming Design Objectives

The stimulants for driving this project forwards are as follows with the four 'NO's' in bold as seen in Fig. 3.1:

- Working from the inside out.

This approach is considering all interior functions as being expressive of the form of the building.

- Person/activity is the starting point.

This is realised in Chap. 4 in terms of space and modular co-ordination.

- **Reducing the square corner's dominance—spiritually unwelcoming.**

The heartbeat of my approach making me feel like David facing Goliath! Contravening history and convention to make things more organic. See Sect. 4.3 for further explanation.

- Developing a cellular concept that reflects on usability of space.

The 'product' design approach in being disciplined and organised.

- **Dogleg (U-shape) stair to replace a straight staircase which should be banned for being unsafe.**

I have never heard of anyone condemning the straight stair in the home apart from the BBC programme mentioned in Sect. 3.1.1.

- All storage/display has a defined place.

This is part of my 'product' design approach that everything has a place.

- Giving storage proper consideration.

Increasing the amount of storage is overdue and will demand more space BUT IT IS NECESSARY!

- **No floor to ceiling glazing:** due to heat loss, feelings of insecurity, unsympathetic to furniture and children. All windows have to have a wall upstand. Unpopular with devotees of a fashion that makes no sense. Homes are supposed to be private so why expose them in this way?
- SECURITY: roller metal shutters over doors and windows should be standard.
- Giving proper consideration to children as they grow up.

The most difficult aspect because children have been ignored.

- Removing old style of living and the 'conditioning' of people.

The success of this is dependent upon marketing to 'sell' new ideas.

Fig. 3.1 Four popular elements of house design today to be eliminated or reduced in influence

- **Replace hinged doors with sliding pocket doors** as they are an unwanted intrusion in a room when left open, and they are problematic for toddlers and anyone carrying a tray.

Another social hurdle to overcome. See Sect. 5.6.

3.1.2 Stairs

On 7th June 2016, the BBC4 showed a programme called 'New hidden killers of the Victorian Home' in which stairs were mentioned as being a primary cause of accidents and deaths with people falling down them. I mentioned this in my first book[1] and recommended that all houses should have a maximum number of five treads with a landing break. The constant trend of retired or elderly people living in the dreaded non-architectural 'bungalow' is a misnomer in the sense that with the elimination of stairs, the accident risk is reduced. The elimination of stairs reduces the chance of some exercise, and if they are built to be safe, then they are essential to promoting a healthier existence. There is nothing worse than identifying the bungalow estate with one's last port of call before reaching the Pearly Gates, by witnessing repetitive trips of ambulances followed by funeral processions.

3.1.3 RIBA Research Paper 'The Way We Live Now' 2012

This research project has been commissioned from Ipsos MORI by the RIBA for the independent Future Homes Commission's investigation into how the UK can build enough of the right kind of homes for modern British households. Here are some extracts in italics that refer to the main feedback issues:

Natural Light

Large windows for natural light, large rooms and high ceilings, typically referred to as 'period features'.

Older properties were perceived as being more characterful than modern ones.

Lounge that Works

Large main living area for eating and socialising, either with members of the household or for entertaining guests.

Whilst participants in this survey expressed a desire for more open-plan arrangements, this desire is misguided due to the fact that their experience has been living like sardines with family conflictingly sharing one space for many activities.

[1]*Interior Design: Theory and Process.* Anthony Sully. London, Bloomsbury. 2012.

Incorporate New Technology

Layouts which take into account technology used within the home.

Participants had an expectation that new-build homes would include suitable storage for technology and layouts which reflected how they use technology.

New housing certainly does not provide the right spaces for this.

Privacy

Private space within the home made an important contribution to participants' Space for private time away from other members of the household.

This is a fundamental point, which causes so much tension in families. It is not only good sound separation that is required but the provision of the right number of spaces for carrying out different activities.

Storage

Long-term and short-term storage for functional items, and for personal possessions people have chosen to keep during their lives.

The lack of storage I consider to be a social crime. The days of basements and attics being provided for storage has long gone, leaving householders to use the garage, if provided, or the spare room. It is as though people of all ages are not allowed to carry their memories with them, which is why car boot sales, and Internet selling sites have become so popular.

Proper Utility Space

Dedicated space for domestic utility tasks, such as, washing, drying and ironing clothes, as well as for storing vacuum cleaners, rubbish bins and recycling.

If a utility space is provided, the space is usually for a washing machine and drier only—nothing else. Proper space is needed for ironing, handling the washing and the provision of a sink. It is also the space for household maintenance.

Adaptability/Flexibility

Options for different layouts, as no consensus was reached for a single standard layout which was preferred across all households and life stages.

This indicates how difficult a house can be for a growing family. Also, children have been largely ignored in the design process.

Welcoming Atmosphere

Emotional considerations can overrule practical considerations when people are choosing their new home.

This is fundamental to my philosophy in this book, that people matter—not architectural egos or cost constraints or local authority strangleholds. It is a welcoming atmosphere that counts.

Limited Comparable Examples

Most participants found it difficult to quantify and compare space in homes, but tended not to be aware of these difficulties—especially if they were first-time buyers.

There are so much conformity, repetitiveness and unimaginative examples in house design that people are unaware of potential variations.

No mention of security.

3.2 Domestic Activities

Let us consider the major activities and movement required for a single working person beginning with the start of the day. For a couple or a family, there will be changes or additions to this list. It is important that designers go through these tasks, which may seem mundane and 'everybody knows them', but taking these for granted means that the designer runs the risk of making assumptions which is a popular failing. Most of these are planned, whilst some occur haphazardly. All of these actions rely on items in storage relevant to the particular needs. The mindset of people involved is of prime consideration of every designer. Families will create an extension of these core activities. I do acknowledge that when dealing with an abstract situation for the average person that no such person exists, so I hope the reader will forgive me in my quest for a summary that is enough for me to show a way forward. I also accept that spaces can ultimately be swapped around or reallocated according to the occupier's requirements.

Prior to seeing these activities, I thought I would show you a page from the undated 'The Housewife's Book' (Fig. 3.2) but appears to be around 1930/40s that lists these daily chores in detail and with recommended times that our superwoman should carry out such 'duties'. Note the heading is for a 'Serventless House' of 3–4 people, which falls in line with my own chosen category later in this chapter. Some curious anomalies are that she does not appear to wash first thing in the morning; the stairs are swept indicating that they are solid wood, and the scullery was the early utility room. Note 'Tea' is taken at around 4.00 p.m. and children are put to bed around 7.00 p.m. I do like the soft-hearted concession that after the evening meal the washing up could be left until the morning! These times were a continuation of the nineteenth century's role of the housebound wife who looked after, cleaned and managed the finances of the home. It is interesting to note how the kitchen, which used to be the messy working back room concealed from visitors, slowly emerged over the decades to be a reception room and hence much thought went in to cleaning it up to such a ridiculous state that belies its true productive nature as what I like to call a 'cooking workshop'. They are now more like 'showroom kitchens'!

Over recent decades the trend has been for both partners of a family to work full-time until children arrive. But even then, due to economic pressure for survival, child care, nursery and playgroups are used to enable the parents to continue their day jobs.

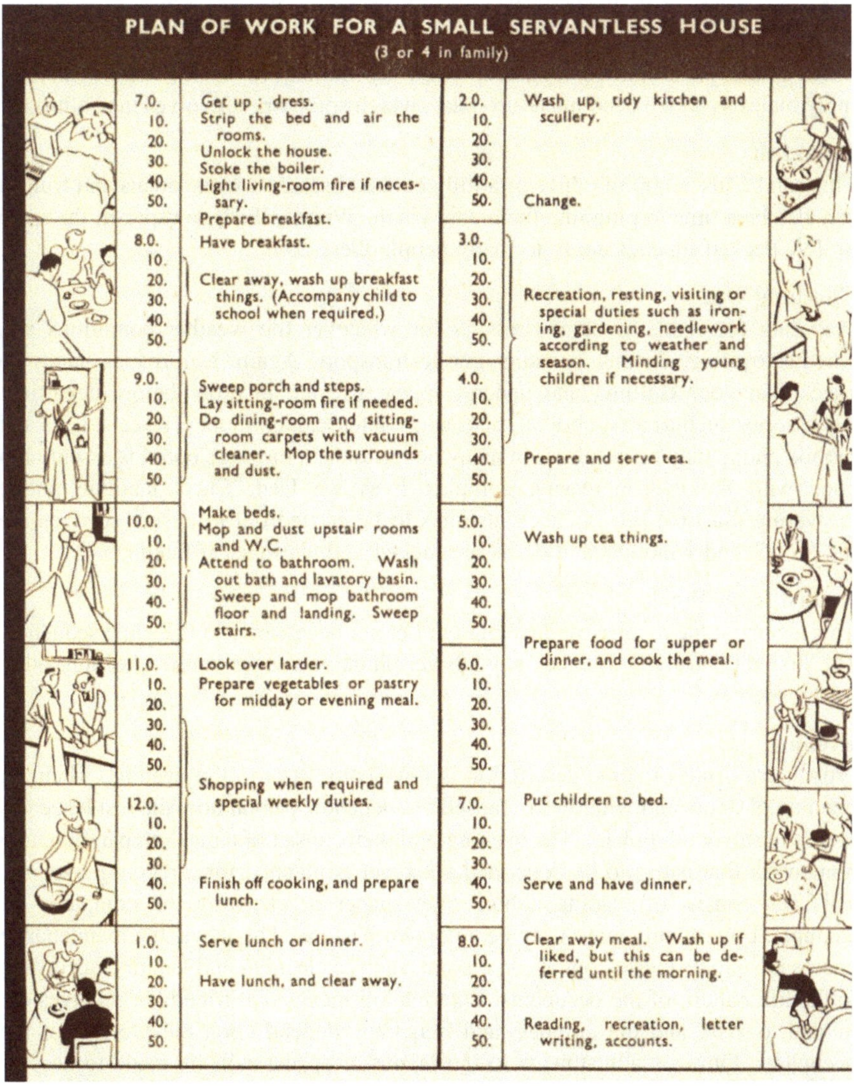

PLAN OF WORK FOR A SMALL SERVANTLESS HOUSE
(3 or 4 in family)

Time	Duties	Time	Duties
7.0. 10. 20. 30. 40. 50.	Get up ; dress. Strip the bed and air the rooms. Unlock the house. Stoke the boiler. Light living-room fire if necessary. Prepare breakfast.	2.0. 10. 20. 30. 40. 50.	Wash up, tidy kitchen and scullery. Change.
8.0. 10. 20. 30. 40. 50.	Have breakfast. Clear away, wash up breakfast things. (Accompany child to school when required.)	3.0. 10. 20. 30. 40. 50.	Recreation, resting, visiting or special duties such as ironing, gardening, needlework according to weather and season. Minding young children if necessary.
9.0. 10. 20. 30. 40. 50.	Sweep porch and steps. Lay sitting-room fire if needed. Do dining-room and sitting-room carpets with vacuum cleaner. Mop the surrounds and dust.	4.0. 10. 20. 30. 40. 50.	Prepare and serve tea.
10.0. 10. 20. 30. 40. 50.	Make beds. Mop and dust upstair rooms and W.C. Attend to bathroom. Wash out bath and lavatory basin. Sweep and mop bathroom floor and landing. Sweep stairs.	5.0. 10. 20. 30. 40. 50.	Wash up tea things. Prepare food for supper or dinner, and cook the meal.
11.0. 10. 20. 30. 40. 50.	Look over larder. Prepare vegetables or pastry for midday or evening meal. Shopping when required and special weekly duties.	6.0. 10. 20. 30. 40. 50.	
12.0. 10. 20. 30. 40. 50.	Finish off cooking, and prepare lunch.	7.0. 10. 20. 30. 40. 50.	Put children to bed. Serve and have dinner.
1.0. 10. 20. 30. 40. 50.	Serve lunch or dinner. Have lunch, and clear away.	8.0. 10. 20. 30. 40. 50.	Clear away meal. Wash up if liked, but this can be deferred until the morning. Reading, recreation, letter writing, accounts.

Fig. 3.2 Daily duties from 'The Housewife's Book' published by the Daily Express. *Courtesy* The Daily Express

3.2.1 Typical Modern Daily Sequence of Activities (Without Children)

I need to emphasise that these sequences are major targets of operation, but they can be bypassed, reversed or something else added according to the individual's desires.

Waking Up

Every person will regain the awakening brain in order to remember what the day has in store for them. Most people rely upon an alarm device to wake them. Use of the toilet is usually the first calling. Washing/showering will assist with this process as looking in the mirror will confirm who you are even if it is not the image you had in mind. The day's tasks will suggest to you what clothing to wear. Females (increasingly males as well!) need make-up space, which sometimes involves the bathroom.

Breakfast
This can be just a cup of coffee or a fully cooked breakfast with cereals and fruit. It can also be a time to plug into the media via the iPhone, iPad, newspaper, the radio or TV. Packed lunches apply to many people these days.

Checking Out
Preparing for work involves dressing for whatever the weather conditions are, whether cycling, driving or using public transport. Again, a mirror is handy to check that your clothing and general appearance are in order before departing. Accessories such as an umbrella, rucksack, briefcase, bag, money, sweets, keys and mobile phone may be required. In many households, a message board is essential to cope with the ever-increasing complex lives we lead. Check landline phone answering machine set. A check also has to be carried out to ensure services are turned off, and windows and doors are locked and alarm set if fitted.

Return from Work
This is almost the reverse of checking out but not necessarily in the same sequence. The first act is usually to check any answer phone messages before checking other media sources.

Evening
Probably a drink of some description is a welcome start to the evening. Then, an opening of doors and windows is carried out depending upon how the residence has access to surrounding land. The evening will then consist of a meal preparation, any homework that needs to be done, and whatever is planned for a relaxing evening. This can consist of reading a book, newspaper or checking the computer for example. Entertaining guests can be a popular pastime. The inevitable maintenance chores have to be carried out at some point, such as cleaning and laundry, according to the life pattern of the occupants. Entertaining means that friends or relatives are invited to visit, and the numbers and frequency depend upon the lifestyle of the occupants. Obvious adjustments to the layout of spaces will be made to accommodate such functions.

Weekend
Saturday morning usually invites the proverbial lie-in because it is not a working day, but what often happens is that the day becomes a day of opportunity in carrying out tasks that were not possible during the week. Leisure and sport pursuits beckon and these will impose a strain on storage requirements in the home. Sunday morning is definitely the 'lie-in' morning, and the day usually unfolds as being the most relaxed day of the weekend centring upon the main meal of the week at lunch or dinner time. Going for walks is a popular British past-time, as well of course continuing with leisure pursuits.

It occurs to me that the above activities do not suggest a type of space or spaces, nor indeed what sort of income group or social strata the person may occupy. If this is the case, then it is possible to carry out all of these in one room accepting that a toilet/washspace would be separate for privacy. This minimal existence would most commonly be contained in flats or prefabricated dwellings. Or, looking at the other extreme, these activities could relate to the owner of a mansion. The difference being is in the lifestyle of the user from the minimal to the excess of space use. The minimal user will not necessarily be a person of limited means but rather someone who thrives upon efficient provision of facilities as being appropriate to the survival path being followed. The mansion owner, however, looks upon the amount of space as reflective of wealth, a source of comfort, views, vistas and gardens as therapy but hardly based upon the efficient needs as is the case with the minimal user. In fact, this owner will have a whole list of additional maintenance duties associated with such an estate and would most probably be carried out by hired staff. The minimal user will have none of these apart from cleaning and refuse collection.

The following diagram in Fig. 3.3 represents the above in terms of linear sequences that will inform the planning process in that space use is suggested. The important point to note is that the sequences are broken by three choice

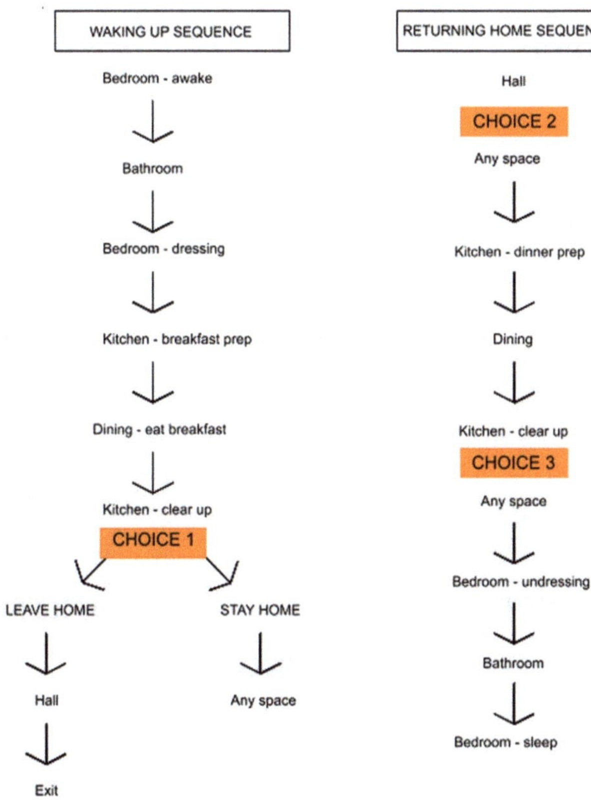

Fig. 3.3 Linear diagram of two most common activity sequences

situations whereby people have to make a decision as to where their next point of call is to be. Of course, in reality random movement can be made as is human nature, but a designer needs some degree of expectation as to the more common movement in order to plan the adjacency and location of space. Existing terminology is used for space names.

3.2.2 *Major Domestic Useable Spaces*

What are the primary activities (space occupying) of the home and how will they relate to each other?

Figure 3.4 shows a bubble diagram of primary activities or functions that can demand their own space. The links between them and relationships or possible

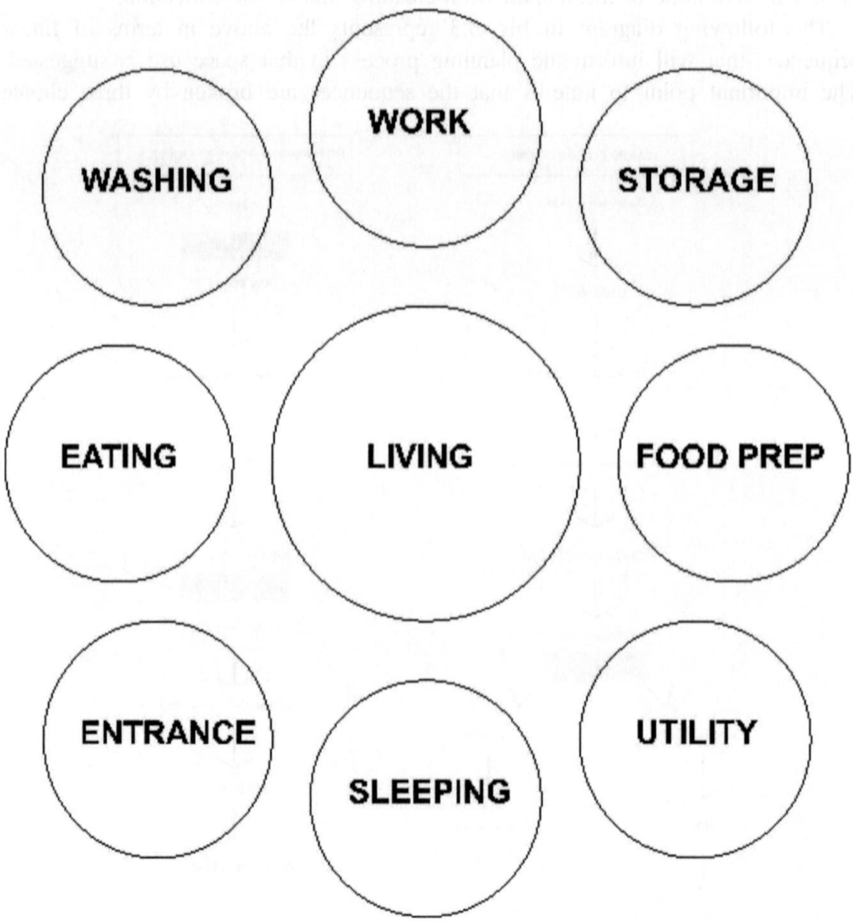

Fig. 3.4 Primary planning activities/functions (storage is shared)

overlaps are not shown as they are dependent upon particular lifestyles although in many instances overlaps can occur whereby eating could merge with many other activities. For example, the 'Dining Room' has become almost obsolete in the modern home as a sole place to dine, and often people will eat in the kitchen or the living room, or anywhere which allows people to eat separately. Many of the case studies in Chap. 2 show a joint space of kitchen/dining/living. To my mind though, whilst this provides flexibility of use, it is denying the proper facility that each deserve. A proper place for dining is almost setting the stage for discussion, stories, humour, games and so on. Shared spaces almost detract from this. They have become fashionable and delusionary to the potential occupier by giving the impression of a 'large' space, which is associated with high status. Interestingly, in the 1950s 64% of women told an Ideal Home Exhibition survey that they wanted the kitchen to be separate from the living space.

Secondary storage may not have a defined space of its own, but it should be a vitally important component of a dwelling as it is the less frequently accessed yet stores valuable items. Unfortunately, it is undermined in the sense that many homes do not have sufficient storage built into the design. Hence, a garage ends up being the main store instead of being used to park the car. The only provision of primary storage, which contains the more constantly accessed items, exists in such areas as bathrooms and kitchens, and sometimes built-in wardrobes. The paraphernalia of goods that families can accumulate over the years has no place. It is as though people are not allowed to have memories. There is no provision for storing shoes, and space for hats and coats is a joke. The fact that people may be entering the house through either the front or rear entrance covered in mud and soaking wet is not catered for, apart from the concessionary doormat. The workspace can sometimes be described as a study, which I accept can be tailored to more specific uses such as hobbies. Why is such a space not considered as important as the main spaces when millions of people derive huge satisfaction from tinkering away with making, repairing or DIY? Why has the larder disappeared? This is a ventilated cool cupboard that fresh produce is stored rather than stuffing it all in the fridge.

The utility area (old name laundry), if not planned as a single space, is another function that is often spread around other areas; for example, the washing machine and drier are often built into the kitchen which is not ideal. Ironing has to be done, and clothes have to be dried somewhere especially in the winter months. Space is required for drying racks, and a large airing cupboard can also help with drying. The utility rooms in new houses are usually so small that ironing has to be done elsewhere. Sleeping usually takes place in bedrooms, but it is common practice for people to nap in living rooms as well. Increasingly over the years, many childrens' bedrooms have become multi-purpose in that they act as study/living spaces. The following is an expanded description of the above activity list in Fig. 3.4:

Living/Lounge	General hub of meeting, conversation, TV etc. Seating for relaxation. TV other media devices. Storage and display. Low table surfaces for magazines or refreshment purposes.
Cooking workshop	Cooking and preparation of food. Cooker. Sink and drainer. Storage for cooking and cleaning. Fridge/freezer. Larder. Preparation surfaces. Storage for preparation tools. Storage of cutlery and crockery.
Dining/Play area/TV	Eating and drinking. Recommend to double up as a child play area. Dining table and chairs. Storage for toys.
Utility	Washing machine and drier. Ironing board and iron. Hoover. Storage for continuous sequence of dirty clothes to washed clothes to ironed clothes. Tools, household maintenance and cleaning.
Bedrooms	Sleeping mainly but could provide a study area. Bed, bedside table, chair, wardrobe, chest of drawers, (dressing table). Some families choose to have an 'extra' bedroom for study/music practice or any use that does require more privacy. Childrens' bedroom almost doubles as a living/study space with TV, computer etc.
Bathroom	Washing and toiletry functions. Bath/shower, washbasin, WC. Storage for cleansing materials. Place for towels.
Separate WC	Required in two storey building on both floors. Plus washbasin.
Workspace	(a) Clean space for any person doing work or hobbies. Desk and chair. For computer, printer, scanner. Writing and reading. Storage for paper and associated stationery. Books, files, etc. (b) Dirtier space for DIY, repairs etc.
Front entrance	Hall. Entrance and exit. Mail delivery. Knocker or bell. Storage for hats, coats, scarves, gloves, shoes, car keys and umbrellas.
Rear entrance	Storage for coats, wellies for garden access.
Airing cupboard	Storage of linen but also for drying clothes.

Note The finished decoration and furnishing (including mirrors) will be left to the occupier. This includes localised lighting fittings, which are separate to the basic provision of lighting that the building will accommodate. It is also important to realise that music can be played anywhere through the use of mobile devices and blue tooth speakers. This affects how spaces are used such as bedrooms whereby they can become study areas as well as areas for relaxation. Building services including alarm systems and telephone/media services will need accommodating at a later stage of this development.

Additions to this list would be a workshop and garden store depending upon the interests of the occupants. These are usually provided by shed type buildings at the rear of the property. The stair/circulation area will be the central hub with all spaces planned around it. All spaces require waste receptacles.

3.2.3 Time to Review Space Names—(See Chap. 5 and My House Design Proposal)

All names/words have derived from other words or meanings, and the change over time keeps pace with social developments. For example:

The Telephone
The telephone made from 'tele' meaning far, distant, and 'phon' meaning sound—this was shortened to 'phone' commonly used as an abbreviation—Mobile phones (cell phone in the USA) was reduced to 'mobile'—then, the iPhone was invented.

The Toilet
'Privy'–'Commode'–'Water Closet' (WC)–'Lavatory'–'The John' (American) 'The Pan'–'Khazi' (from Italian)–'Loo' (from French)–'Toilet'. Also linked with the bathroom when toilets were sharing the bathing space instead of being separate.

The Home
All the terms that we use to describe the rooms in our homes have been around for centuries, depending upon the status of the family, apart from such terms as 'Chamber' for room. This is confirmation of the lack of design ingenuity when no new words appear.

The assumption here is that for a three-bedroomed house the main serving spaces (living, kitchen, dining, etc.) are on the ground floor and the sleeping areas are on the first floor (see Sect. 5.1.3). We have inherited over the years the names of the rooms in our houses almost without question. But due to the changes in modern living, it is now time to revisit these and question their validity. There will be some people who will react by saying that tradition and familiarity are good soulmates so why change? Any design sequence involves a full investigation as to why and how we have come to this situation. I am well aware that changes usually take place organically and are encouraged by journalistic observations, so I accept that my proposals may need time for acceptance. My aim is to achieve one word to describe a space (except cooking workshop) or retain a name if is strong enough. For example the word 'room' is superfluous as it is quite obvious that they are nothing else but rooms. Let us examine the following:

Out of fashion or for wealthier houses	Traditional	Recent additions or changes	New proposed
Vestibule, hallway	Reception hall		Hall
Drawing room, Salon, parlour, front room	Lounge, living room, Sitting room, reception room	TV room	Arena
Dining room	Dining room		Hub
Morning room			
Pantry	Kitchen		Cooking workshop
		Living, dining, kitchen	Not recommended
	Breakfast room	Combined with kitchen	Not recommended
Scullery/laundry	Utility		Utility
	Cloakroom		Toilet
		Conservatory	Not essential
Library	Study	Office	Workspace
Boudoir	Bedroom		Bedsitter
Playroom		Join with dining	Hub
Box room		Store	Store
Washroom	Bathroom	Shower room, ensuite	Cleansor
Games room	Games room	Gym	
Garage	Garage	Multi-purpose	Den

The following are oddballs in that they fall somewhere between room and cupboard and are hardly used today: a pantry is a walk-in area to support a kitchen; a closet is a walk-in cupboard; and a larder is a cool ventilated cupboard to keep fresh produce usually on marble shelves.

The living room or lounge seems rather more difficult to rename due to the multi-functional use of the space especially when merged with the kitchen and dining areas. 'Living' is quite an absurd name as it already has a wide use and does nothing for clarity of purpose. And 'lounge' refers to a person's relaxed body posture, neither of which seem ideal. Estate agents seem fond to describe the main activity rooms on the ground floor as 'reception' rooms when in fact the word reception is hardly used in common parlance. It also has associations with an office. How about 'Arena' which is loosely linked to the entertainment aspect? With TV becoming so dominant, some families may prefer to designate one space as a TV room in addition to a living space—not uncommon these days (see Fig. 5.67).

The dining room has been popularly merged with the kitchen creating the 'kitchen/diner', but also the living room has been added to create an open-planned space with no name. My proposal is to not have the open-planned arrangement but rather merge dining with play area because the space is used for so much drink/snack/conversation occasions, and toddlers need to be close to a parent if busy cooking or washing clothes and in a safe contained space. The open-planned space is definitely not toddler-friendly and hence creates huge problems for the supervising parent. I propose the word 'Hub' as this could be the hub of the house where occupation of the space is most frequent over a working week.

The separate toilet on the ground floor is often off the hall (worst place it could be) and also used for hanging coats (under the stair cupboard another option) as there is no other proper space allocated. Hence, it is called 'cloaks' short for cloakroom. Who wears cloaks these days? This again does not work when coat access can conflict with someone using the toilet. These house designers also forget how much coats hung on coat hooks can stick out from the wall reducing the narrow space allocated for the toilet and washbasin. Assuming the toilet is planned elsewhere and is not used as a coat hanging place, I propose to simply call it a 'Toilet'.

The bathroom, which frequently has a bath, toilet and basin and sometimes a shower facility, is also too old fashioned a term. Although some bathrooms have showers and no bath hence the loose application of this term. I did consider the term 'washroom' but that has connotations with washing clothes. This room is essentially about cleansing the body or part thereof. I therefore propose a word that derives from cleansing but does not exist, namely 'Cleansor'. I will also repeat what I have said in my book 'Interior Design: Theory and Process' that men should sit on the loo to pee just like a woman for health reasons in reducing urine spray when standing. Or install a wall urinal. Toilets appear much cleaner as a result according to my own practical experience.

The study (or office) has become much more popular in recent years and can appear on the ground or first floor, but the use of this space has not been clearly defined. Initially allocated for one senior member of the household combined with some aspect of working from home, both parents and children now can all use such media devices as the mobile phone, iPad, laptop or desktop computer for working purposes. In current homes, the portable devices are used in any room of the house! (see Sect. 3.2.6). This is an absurd denial by designers in not creating specially allocated workspaces, which I have done in Chap. 5. My proposed word for this space is 'workspace'.

The bedroom is another space that desperately needs its name changing. It has become multi-purpose in that adults or children are tending to combine other activities with sleeping because their busy lives are not accommodated in the ground floor serving spaces. Bedsitting room is the obvious choice, but in order to rid myself of the word 'room' and not have two words, I have decided to use 'bedsitter'.

Lastly, the Garage (not many people put their cars in the garage) has to be included even though it is not officially an interior space it has become over the years a store, a utility area, a workshop or anything which cannot be done in the house. The frequency of such use depends upon the proximity and access to the garage. My suggestion for my house design project in Chap. 5 is to call it a 'Den'. It can be a place of creativity and enjoyment where messy work can be undertaken without fear of spoiling the house. Also, the word 'garage' refers to a commercial place for repairing cars, or it can be a depot for buses, so its continued use as part of a home is past its 'sell by' date. A carport is a different term to a garage because it is not an interior space and only offers a car shelter from rain and snow. Its origin is from the french word porte-cochère meaning a covered portal.

3.2.4 The Garage (See Sect. 5.3.7)—The Den

In the many houses shown in this book so far, a garage has become an inconsistent element compared with other provisions that make up a house (see Sect. 1.2). Even now some new housing estates do not incorporate a garage but rely on ensuring that there is adequate parking space for two cars. James Wentling[2] illustrates the various relationships that garages have had to the main house from being detached to being attached or even integrated. He also laments how it came from being of secondary importance architecturally in the past to being more assertive at the front of the house resulting in appalling streetscapes (Figs. 3.5 and 3.6).

Figure 3.7 shows a popular suburban street that does not have enough room for cars to be parked on both sides of the road. The old pretty front garden has been replaced by an intrusive machine sitting on hardstanding at the hands of an uncaring house owner.

Fig. 3.5 Detached double garage too dominant. *Photograph* Author

[2]James Wentling, *Designing a Place called Home*. Switzerland, Springer. 2017. p 76.

Fig. 3.6 Integral double garage too dominant and out of scale with the rest of the house. *Photograph* Author

Fig. 3.7 Front gardens of these 1930s semis making way for the car. *Photograph* Author

3.2.5 Interior Accommodation

My main argument is that the interior accommodation provided is generally inadequate on many levels and this book attempts to highlight this by developing a design approach that is intended to show a way forward. When I mention the interiors of houses, the general perception seems to be of the decoration, the lighting and the kitchen and bathroom fittings. My focus in this book is on the form, shape and facilities provision of the interior spaces and their relationships and not the decoration. Here are two extracts from the Government's paper on 'Design and Quality Standards' 2007:

(1) **Internal Environment**
Internal environments should be comfortable, convenient, capable of sensibly accommodating the necessary furniture and equipment associated with specific room activities and be suitable for the particular needs of intended user groups.

It is quite clear from current experience that this aim is not being fulfilled, and one wonders what they mean by 'sensibly'. Likewise, it would have been more helpful if they gave examples of the type of 'furniture and equipment' as this does not increase our expectations to a better provision.

(2) **Sustainability**
New homes should be designed and constructed in a sustainable manner using products and processes that reduce environmental impact, better adapt to climate change, with lower running costs and incorporating features that enhance the health and well-being of constructors, occupiers and the wider community.

A rather woolly statement but one which is very much being acted upon in terms of prefabricated factory-made components and energy-saving insulation and power generation devices. The following quotation focuses on feedback from the public which is all to revealing:

> *Today's home buyers are dissatisfied with the new homes available to them. Too few reflect the requirements which matter to would-be home owners: Good sized rooms, storage, natural light, privacy and good sound insulation, and the flexibility to accommodate future needs and family members. Poor design means few communities welcome the prospect of new homes being built nearby, so developers are involved in "constant local battles". The Future Homes Commission: Building the Homes and Communities that Britain Needs*[3]

'Good sized rooms' is mentioned above and of course this refers to how small new houses are in the UK compared with those built on the continent. You will have found in my appraisals of the case studies many such criticisms. Comparison with the USA is not made easy when they have so much more space available as shown by the World Bank statistics of nos. of people per km^2: USA 35 and the UK 269. So our small island

[3]*The Future Homes Commission* (October 2012).

reflects this in car parking spaces. The width of a space in the UK is 2.4 m, and in the USA, it is 2.8 m wide. In British car parks, it is sometimes difficult to open the door enough to get out!

3.2.6 *Influence of Media Communication Devices*

This development has accelerated the break-up of families in that they commune less with each other than in the past. When I was a boy, we had no TV and a visit to the cinema was a huge occasion that used to happen about once a fortnight. Today, it is possible to watch a film (or more than one) every day one way or another. This glut of supply does nothing to increase social engagement. Back in the 1950s, there would be a local pub serving every village or town district whereby neighbours of both sexes would meet regularly to gossip and put the world to rights. This separatist trend will continue with the advanced use of laptops and iPhones. It is astonishing to witness how many people, at any time of the day, can be glued to the hypnotic touch control of these tiny screens. They have accessibility and usage 24 h a day, which not only impinges on other activities, but expands the knowledge base of people to a highly informative level. I have not seen one house plan that responds to modern living in terms of accommodating the increased use of IT and media devices. The fact that many such devices are portable does not help the situation in that a laptop can be used whilst eating, whilst watching TV, or in bed! The notion of providing one 'study' area which is usually reserved for the senior worker of the household again is outmoded. How many times can one witness when a normal discussion is taking place, a question is raised, and in the face of no answer being forthcoming, referral is made to the instant God of communications—the iPhone. The owner of which stands proud as though he/she has solved the problems of the universe. The following is an extract from an article published in the Guardian warning of the resulting effects social media is having:

> *Twenty-five years went online, it is clear that we are living through a period of dizzying transition. For 500 years after Gutenberg, the dominant form of information was the printed page: knowledge was primarily delivered in a fixed format, one that encouraged readers to believe in stable and settled truths.*
>
> *Now, we are caught in a series of confusing battles between opposing forces: between truth and falsehood, fact and rumour, kindness and cruelty; between the few and the many, the connected and the alienated; as its architects envisioned it and the gated enclosures of Facebook and other social networks; between an informed public and a misguided mob.*
> Katherine Viner[4]

Figure 3.8 shows up how design has become so slimmed down with the iPhone as to be an insignificant object by comparison with the old telephone. Of course, there is a wealth of differences in function: the telephone was just a telephone, whereas the iPhone is a small computer storing huge amounts of information,

[4]Katherine Viner, Guardian 14.7.16.

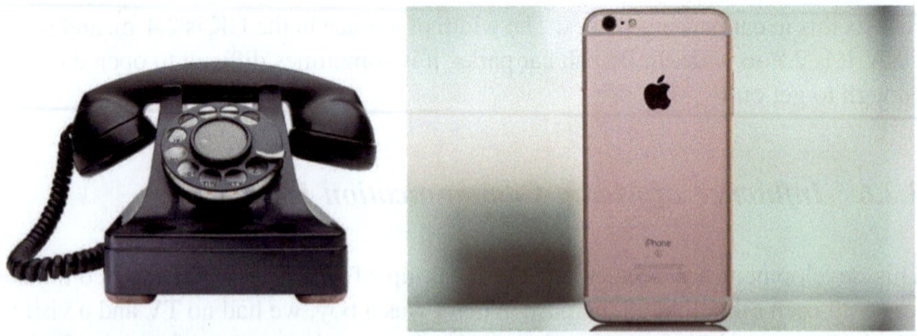

Fig. 3.8 Popular old telephone from 1930s on left contrasts with the modern Apple iPhone

Internet functions, a torch, as well as being a camera! Miniaturisation is the wonder of the digital revolution but my goodness, how boring as an aesthetic object! It reminds me of when the cardboard cigarette packet was discarded for a sleek chrome/silver case just as thin as a cigarette, and not unlike the mobile phone. Usually, when functions increase, there is a comparable increase in the size and capacity of the object, but not in this case. Ergonomically, it is a failure: it is difficult to hold and operate. Whilst thought has been given on its functions, no thought has been given on making it easier to hold. Despite its immense value and importance, it is treated with disdain by many people when it comes to carrying the phone on their person. It is put into the back pocket of their jeans or some other pocket without any special consideration. They often place their phone on a bar counter or table for all to see and often complain that they have lost their phone or cannot find it. I myself put my phone in a leather pouch, which is attached to my belt and to my amazement, despite broadcasting the fact that it is impossible for me to lose my phone, I have not met anyone else including my whole family who does the same. The belt appendage attracts derision, but this is to me is an anomaly in human behaviour. The old telephone was tied to a location by cable and was not mobile. People had to travel some distance in order to make a call, and they would not know if anyone was trying to contact them until they returned to their home or workplace. The mobile phone has enabled people to communicate 24/7 anywhere, which has created a safer society, a speedier society in all kinds of dealings and a trackable society. It has also paved the way for the unscrupulous criminal element and anti-social behaviour.

3.3 Children

3.3.1 Children Ignored

The impact on a couple of having a child can be a combination of delight, trauma and disturbance to what was a repetitive routine of activities. It maybe sometime, especially in this climate of inflated house prices, when a young couple will be able to afford a house, and instead their first family home will invariably be an apartment. However, there are many older couples also having children who will be living in a house of some type. Whatever the situation, I see no evidence of house design properly considering children. It is as though they do not exist. A 'playroom' used to be quite a common feature in a home but that seems to have disappeared over the years as children's bedrooms have become more like bedsitting rooms. These little aliens are constantly being told off for touching things they shouldn't and subsequently cause unhappiness for them, and stress for the parent. Bathrooms are designed for adults—not children. I have suggested in Sect. 5.3.1 that the dining area could double up as a play area with proper toy storage.

3.3.2 Child Safety

The following diagram is from the website of the USA company called Safe Kids, and it seems to mention the obvious but also missing a great deal. The statistics of fatalities is shocking (Fig. 3.9).

There are five of these sources listed above as follows:

Fig. 3.9 Major sources of accidents to children in the home. *Courtesy* Safe Kids Worldwide

1. Location of medicines—must be out of reach.
2. Cot suffocation.
3. Bath incidents.
4. Smoke alarms not checked.
5. Furniture/products tipping over.

1, 2 and 3 are down to parental care, whereas the only two related to the design of the house are 4 and 5. In addition to these points, there is also advice about the dangers of the kitchen and straight stairs which basically means physical prevention in the kitchen and childproof gates across the stairs. Designers have a duty to identify all potential hazards by not providing them in the first place. As Fig. 3.1 clearly shows, there are two hazards that apply to children and that is the straight stair to be replaced by a dogleg stair and hinged doors to be replaced by pocket doors. I have applied a design solution to solve these issues, but more thought is required throughout the interior of the house.

3.3.3 The Growing Child

The grow chair in Fig. 3.10 is a superb example of a product designed to be adjusted for growth of a child. This is the kind of thinking that needs to be applied to house design. Figure 3.11 shows a garment designed by Ryan Mario Yasin a recent graduate of the Royal College of Art, that has a pleat system that lets garments stretch to fit even after growth spurts.

The conceptual drawing in Fig. 3.12 shows the kind of height adjustment that is necessary for any wall-mounted exposed surface that may have valuable products that need to be kept away from a toddler's grasp. This is accepting that in a toddler-free environment, there is no restriction on the positioning of furniture surfaces. This excludes enclosed furniture items as long as the doors are child proof. This principle needs to be applied to all spaces that have such vulnerability. Once the child has grown to a more discerning and responsible age, normality can return. The coffee table in the centre of such spaces is also a prime target of the toddler but is also the cause of many tears when fallen against it. My recommendation is that

Fig. 3.10 Tripp Trapp grow chair designed by Peter Opsvik in 1972. *Courtesy* The designer

Fig. 3.11 Expanding garment designed by Ryan Mario Yasin, 2017

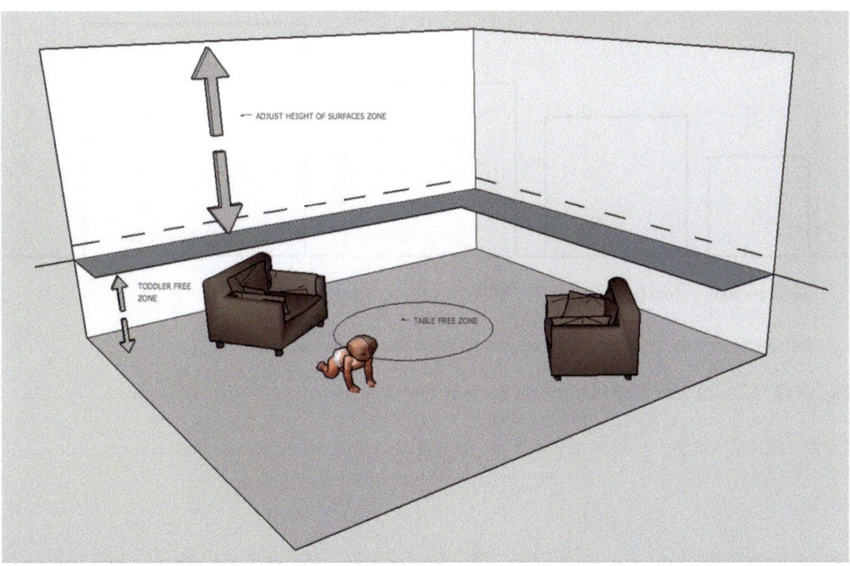

Fig. 3.12 Living space adjustment for toddlers

alternative solutions need to be sought that creates a clear open space in the centre. Please see the plan in my design scheme in Fig. 5.15 where this is applied.

3.3.4 *Change of House Size According to Family Growth*

From my own experience, the following diagram illustrates how the size of a dwelling responds in time in two ways according to the numbers of people in occupation and their status (Fig. 3.13).

Of course, the above diagram is a generalisation, but I produce it to show that the real space needs of a family are very important because the responsibility of raising a child is huge and house design almost ignores that process. It also shows that the wealthier people do not really need that space in the same way. The Ministry of Housing, Communities and Local Government's report: English Housing Survey, 2014–15, states people aged 65 and above currently have almost twice as much 'living space' as 16–44 year olds.

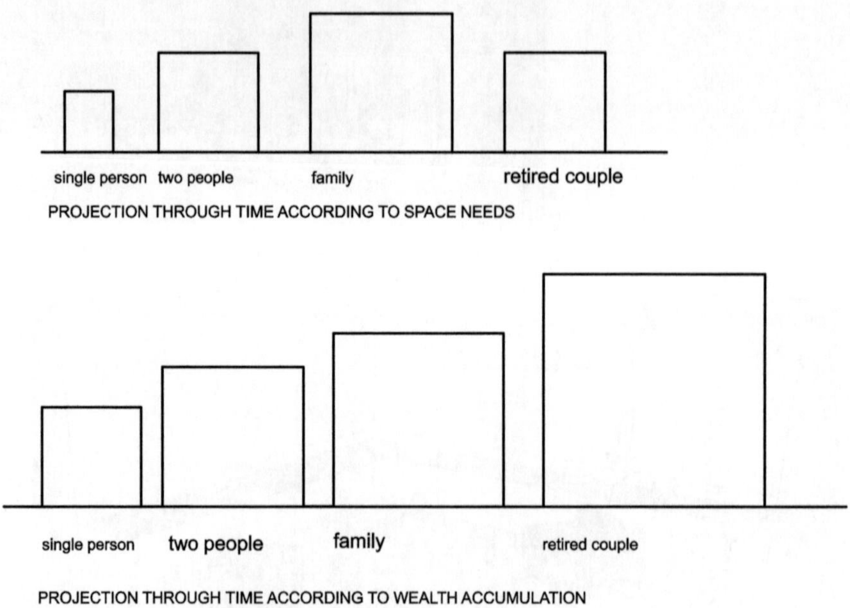

Fig. 3.13 Size of dwelling dependent upon needs or status

3.4 Conclusion

The function of this chapter is to prepare the way in the next chapter for a geometric analysis of the space and layout required by summarising the activities taking place in the home. I have made strong suggestions for change in order to improve house design, and I have examined in detail these activities ensuring that this chapter forms a sourcing brief. The reader may well feel inclined to add more activities or facilities that may be missing and that is an acceptable part of the purpose of this book. The question of accommodating children is one that usually falls upon the parents in that they assess a house in terms of sleeping accommodation only, accepting that an outdoor space is essential. My point being that the design, planning and basic equipping of current new-build houses will not have any features that are child centred.

The depressing thing about many reports that are made about housing is the absence of any reference to the 'design' of the house but more about supply and demand, master planning possibilities and budget requirements. Major problems about space standards usually involve debating the number of bedrooms as an indicator of size and status, whereas the space relationships and their storage provision are glossed over. The common perception for distinguishing house sizes is usually defined by the number of bedrooms, and the main spaces of living, dining and kitchen increase in size proportionately. This is ludicrous! A four-bed house will require the same main spaces as a two-bed house because all the activities carried out are the same and therefore require the same size of space. The notion that a house with a 'spare' room has larger provision is so antiquated. Any increase in the size of a house needs to be dispersed throughout all the spaces. It is no use having a spare room if the bedrooms are undersized.

> *Today rooms are sometimes made smaller to accommodate the prized little strip of garden around the home. Furthermore, because the British price housing based on the number of rooms, rather than the total floor space, a little extra profit can always be made by making rooms in a new property just a fraction smaller.* Danny Dorling[5]

[5]*All that is Solid.* Danny Dorling. UK, Penguin Books. 2014.

Chapter 4
Geometry and Space

Abstract Before any design project begins, it is important to develop a strategy of approach that uses research material directed towards the subject. The previous chapters have prepared the groundwork in research and history, and it is now time to look more closely at the kind of geometry and system of space control that will suit the subject in hand: namely an estate house for mass production. Delineation of space occupation by one person is the beginning, and from that point, a description of areas associated with a particular activity will begin to take shape. The chosen route is organic in defeat of the prominence of the square, and polygonal forms emerge offering a more and human-centred alternative approach. The design process is morphological and organic as opposed to artificially imposing a desired form from the past or some other source.

4.1 Occupation of Space

4.1.1 One Person

A person requires a certain amount of space depending upon the activity or purpose of that space, that comfortably allows basic physical and unrestricted movement. A minimal example could be the sentry box in Fig. 4.1 that is designed as a sheltered space for a member of the Queen's Guards in London. This portable enclosure provides temporary shelter for standing in only.

The plan figure drawing in Fig. 4.2 shows an area of 0.810 mm^2 which suggests a domain prescribed for that person only.

The usefulness of assessing a single person standing space occupation is that it helps to regulate a module for planning purposes. I shall choose 970 mm^2 as my planning module as it encompasses a common width for doorways, although the Metric Handbook[1] determines 750 mm as a minimal clearance space for one person. Another determinant in arriving at a modular grid can be based upon the standard width availability of construction materials such as panels, frames. In this

[1]David Adler, *Metric Handbook*, Oxford, Architectural Press, 1968.

A. Sully, *The Estate House Re-designed*,
https://doi.org/10.1007/978-3-319-90397-2_4

Fig. 4.1 Sentry box, London. *Photograph* Author

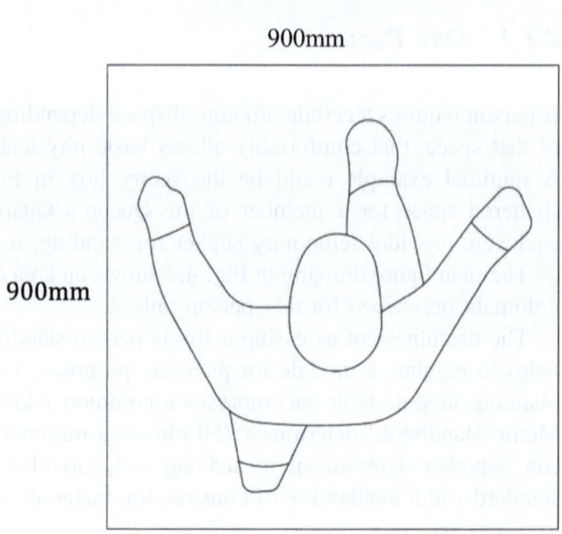

Fig. 4.2 Domain for single person

case, modules of 500 or 600 mm can be favoured in the building industry. The aim is to arrive at small useable areas of space rather than have an excess of space as witnessed in higher cost homes. My 970-mm module will be used as a basis to incorporate within it door and circulation width, seat depth, worktop depth and so on. The actual finished depth will be set by the ergonomic needs of that activity.

So, I am generating a planning concept beginning with a person plus useable space. This will represent an activity in the house and from that a cellular shape will be formed. These cells will then be planned according to those relationships until a plan is produced. From this, a building form will grow and the architectural properties will be modelled. This process is in contrast to most architectural methods which usually stems from the planning of ACCEPTED STANDARD AREAS of each activity inside a pre-meditated building envelope that imposes upon a room its shape, and window and door positions without any detailed concern for the ergonomic or well-being needs of that space.

4.2 Module Progression

4.2.1 Useable Space

From a standing position, a person will choose to walk to a useable space which can be for sitting and relaxing, or it can be a work area such as a desk or food preparation surface, or it may be a storage area. In other words a CONTAINING FUNCTION. So apart from circulation space, there will be provision for an entrance–exit point. This forms the beginning of my planning for the HOUSE in the search for a basic cellular space that can be repeated for a variety of functions (Fig. 4.3).

Figure 4.4 is the starting-off point whereby the planning begins with a standing person showing potential options for expansion in a 360° surround. Useable space could theoretically be all around the person but this arrangement is more direct leaving the corners as choice of entrance–exit. This breaks with the convention of assuming that planning begins with a front door, hallway and reception rooms off as a starting-off point. I was always amused by students when explaining their project plans to me they would begin with ‘You enter here….’ being the front door. Where, as I explained, it is far more important to communicate the overall planning strategy before descending into entrance–exits points. Plan first by spatial relationships rather than by the circulation path. The reason for this is easy to understand simply because the beginning of a journey when visiting a house is the approach to the front door, and the ‘plan’ of the house is never seen. So as I said in my Introduction, I am developing a concept from the inside out. This is not an uncommon approach in architecture as the firm M Moser Associates[2] declare on their Website:

[2]M Moser Associates architects in London, USA and Asia.

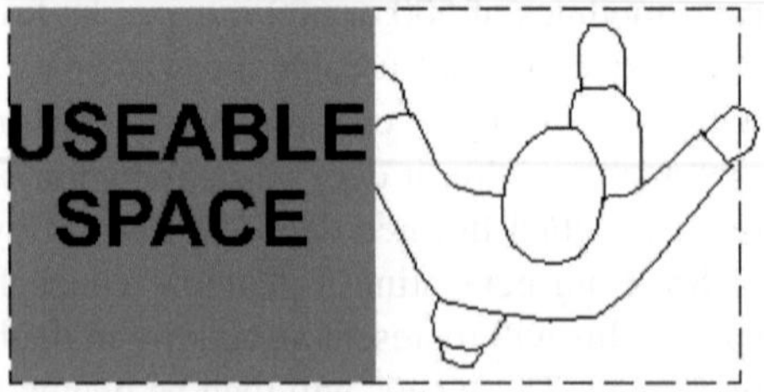

Fig. 4.3 Two modules showing person standing space and USEABLE SPACE

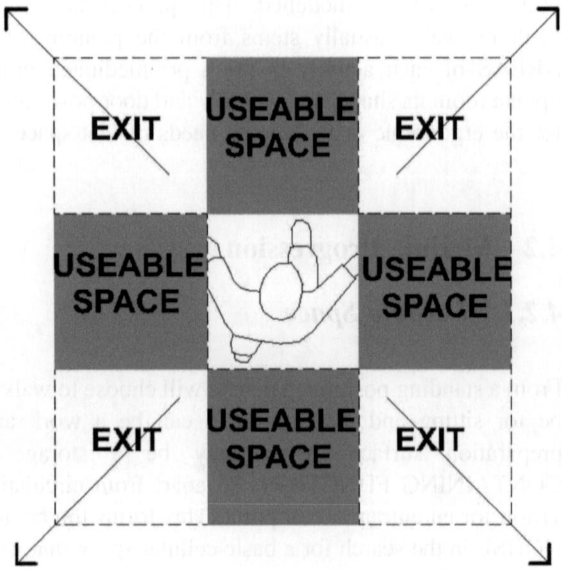

Fig. 4.4 First plan diagram showing immediate module cells to the single person

How a building is designed for a specific end-user has as much impact on its final form as why it was designed. Taking the traditional route of designing a building from the 'outside-in', for example, often results in exterior architecture being optimised to the detriment of interior functionality. For businesses, the physical architecture of these conventionally designed buildings can actually obstruct the operation of a workplace, and therefore impose inherent limits on what strategic goals can be achieved. Conversely, architecture can help remove these limits if it designed to support your goals from the start. This is the essence of Inside-out, an approach to conceiving and designing buildings around the core goals, end-user needs and technical requirements of their users. The result: Buildings that are more accurately fit to short- and long-term purposes, delivered more efficiently, and providing greater value than their traditionally designed equivalents.

4.2.2 Add Entrance and Exit

The difference with my approach is that it is focusing on domestic design as opposed to Moser's project, which deals with the complex issues of corporate design of the workplace.

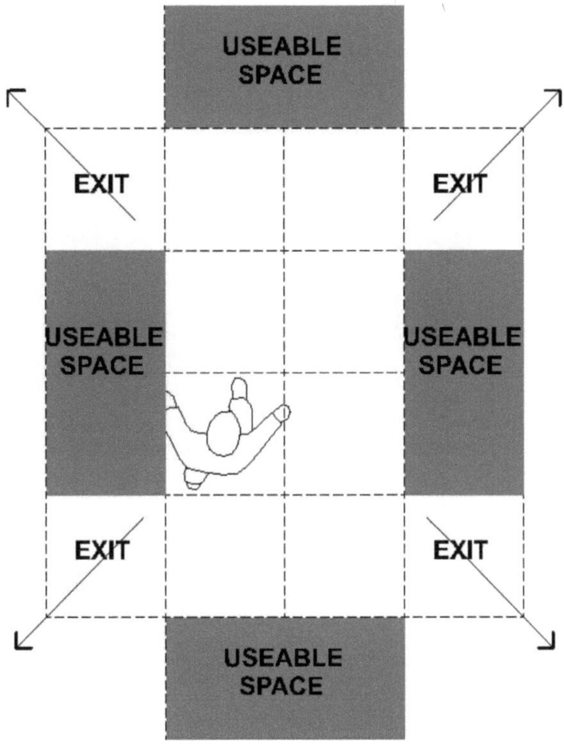

Fig. 4.5 Expansion of space to facilitate circulation

Figure 4.5 shows more space than Fig. 4.4 by providing improved access to double-module (1800 mm) useable areas and single-module exit points. The purpose in this development is to arrive at a useable cell that can be repeated to accommodate all the basic activities in a home. The overall confineable shape in this illustration is rectangular which, when used as a modular repeat cell, has limitations compared with that of a square format.

Figure 4.6 shows a move towards a symmetrical square format using the same double-module useable spaces thus creating further useable spaces in the centre consisting of four module spaces. All of these diagrams are displaying potential placements.

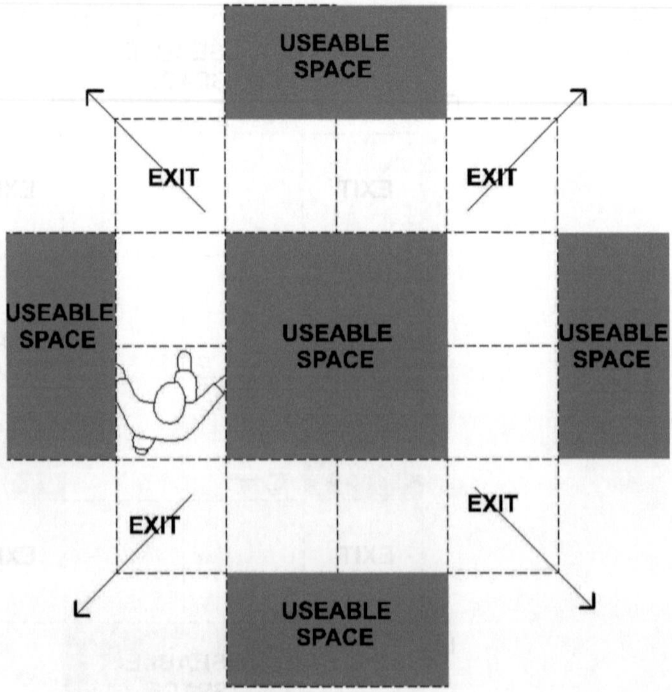

Fig. 4.6 Towards a square cellular space

4.3 The Square

Earlier in this chapter I examined how a single person's workable space can expand according to the particular demands of the activity. I will now regulate this with a geometric override that relates to the ergonomics of space. From my earlier books, the reader will have seen my thoughts about reducing the dominance in planning of 'the box' by elimination of the 'corner' and being one of my 4 No's in Fig. 3.1.

Figure 4.7 shows a segmented square with the external directional pull of the corner 'arrow' combined with the inward-looking arrows pointing towards to the corner. Now if one takes the two adjoining walls and splays them outwards and repeats within a circle, there are two cellular shapes that are popular to work within design and architecture: namely, the hexagon and the octagon. This results in the 'chamfered' corner in order to create a more welcoming and friendlier space and one which I shall pursue in the next chapter. It also acknowledges my aim of reducing the suffocating dominance of the square.

The hexagon has a repeated internal angle of 120° and consists of six equilateral triangles, whereas the octagon's internal repeated angle is 135° and it has eight isosceles triangles .

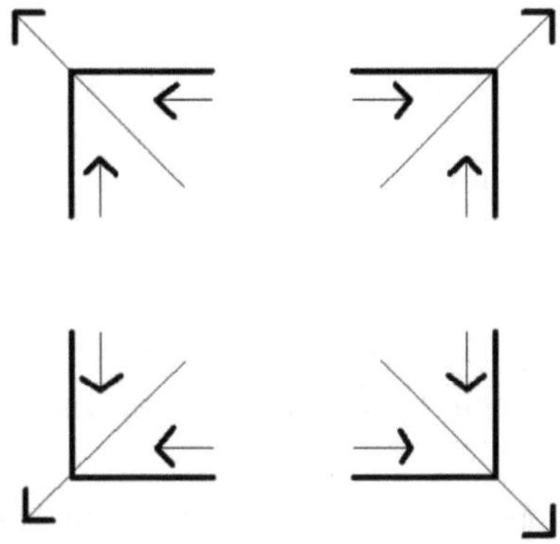

Fig. 4.7 Unwelcome corners

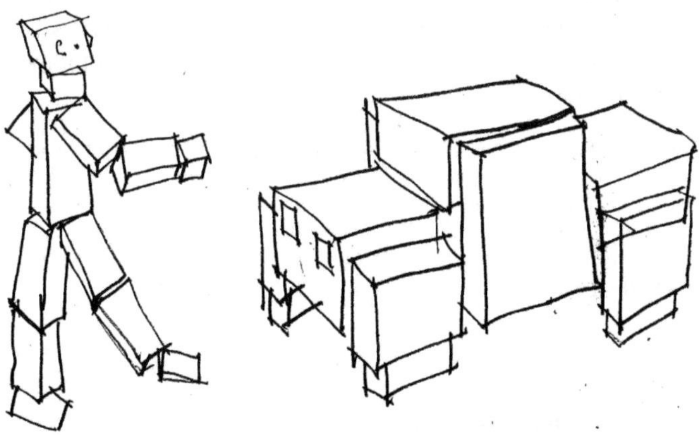

Fig. 4.8 We just are not a boxy world!

Briefly stated, an octagon is a more efficient shape than a square or a rectangle. The argument of course extends to the three dimensional form of these shapes. This is true both in terms of materials used to construct a building and the amount of volume that needs to be heated and cooled with approximately 25% more wall area for a rectangle than for an octagon. It is simple math—surface area to volume enclosed. Of course the shape doesn't have to be an octagon—a hexagon, for example, is also more efficient than the ubiquitous rectangle.

Our love affair with the traditional rectangular shape for buildings means we pay added environmental and energy costs not once, but four times over the life of the structure. First, we pay in the construction phase—more building material means higher "embodied energy" in the building. That is a significant portion of the total lifecycle energy. We also pay for the added surface area with more fuel used to heat and cool a building than would be needed with less wasteful shapes. Greater exposed surface area also means greater maintenance and repair on the exterior (and potentially more damage from storms, hurricanes or tornados than a more efficient design). Allen Dusault[3]

[3] Allen Dusault, *Why Building Geometry Matters.* 2008.

There is a natural explanation of 'our love affair' with the square or the straight line and this is because it confirms the work of humankind as compared with the organic forms of nature. The square represents efficiency and ease of factory reproduction but it can also be boring.

4.3.1 Why Are Most Buildings Rectangular?—Paper Written by Philip Steadman[4] 2006

This is a supportive paper to my argument against the dominance of the square in architecture.

Steadman refers to standardisation in housing, which implied that rectangular components would be used because of ease of repetition and adjacent connections. Otherwise, the continuing popularity of rectilinear planning he states is a mystery. He also refers historically to early primitive dwellings that do not follow the square but were more organic.

> *What all this evidence indicates, I would suggest, is that rectangularity in buildings in the horizontal plane has to do, crucially, with the packing together of rooms in plan. When many rooms of similar or varying sizes are fitted together so as to pack without interstices, it is there that rectangularity is found. Nonrectangular shapes occur on the edges of plans, or in single room buildings, since in both cases the exigencies of close packing do not apply.* Philip Steadman

4.3.2 Chettle House, Dorset, UK

A red-brick manor house in Queen Anne style designed by architect Thomas Archer for George Chafin in 1710. I am showing this house because it is rare to see the dominance of the square reduced in building form to emphasise curved corners. This was a feature of Thomas Archer's work (Fig. 4.9).

The architect chose to soften the building's exterior by curving the corners. Unfortunately, as far as the interiors were concerned the only space that continued this 'force majeure' was the projecting entrance hall and its basement and first floor room. The rooms behind the external curved corners were only incidental as the curved wall bore no relation to the function of those rooms (Fig. 4.10).

[4]Philip Steadman is Professor of Urban and Built-Form Studies at UCL. He works on the relationship of energy use both to the forms of buildings, and to land-use patterns and transport networks in cities. Author of many books including *the Evolution of Designs: Biological Analogy in Architecture and the Applied Arts, 2008.*

Fig. 4.9 West Hall elevation of Chettle House, Dorset, UK. Wikimedia Commons. *Source* geography.org.uk. Chris Downer

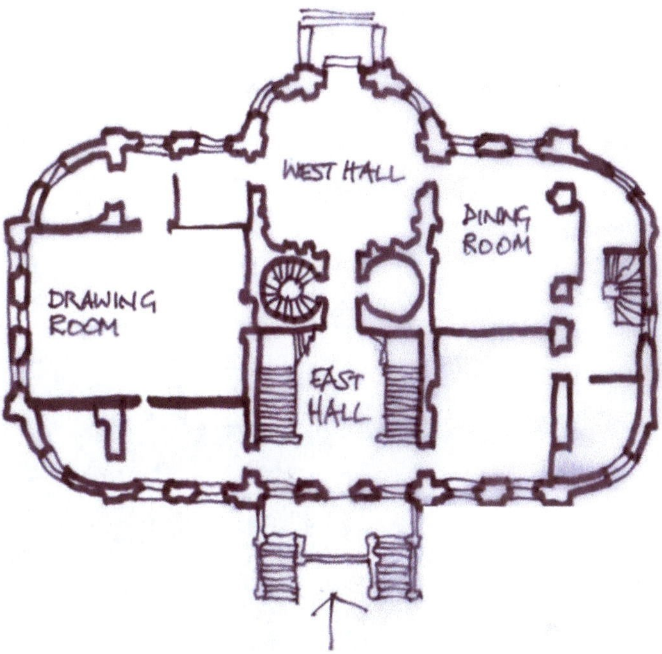

Fig. 4.10 Plan of Chettle House. *Drawn* Author

4.3.3 *Chateau de Monte Cristo, Port-Marly. Yvelines, France*

The château was built in 1846 by the architect Hippolyte Durand the country home of the writer Alexandre Dumas (Fig. 4.11).

This chateau shows the strong modelling of part of an octagon for the spiral stair towers either side of the entrance, as well as the projecting bay window of the Boudoir at the rear. It seems so disappointing that the front elevation reveals such inviting undulation, which is not realised in the main interior rooms (Fig. 4.12).

Fig. 4.11 Exterior view. Both images by permission of Frédérique Lurol, Director of the Chateau

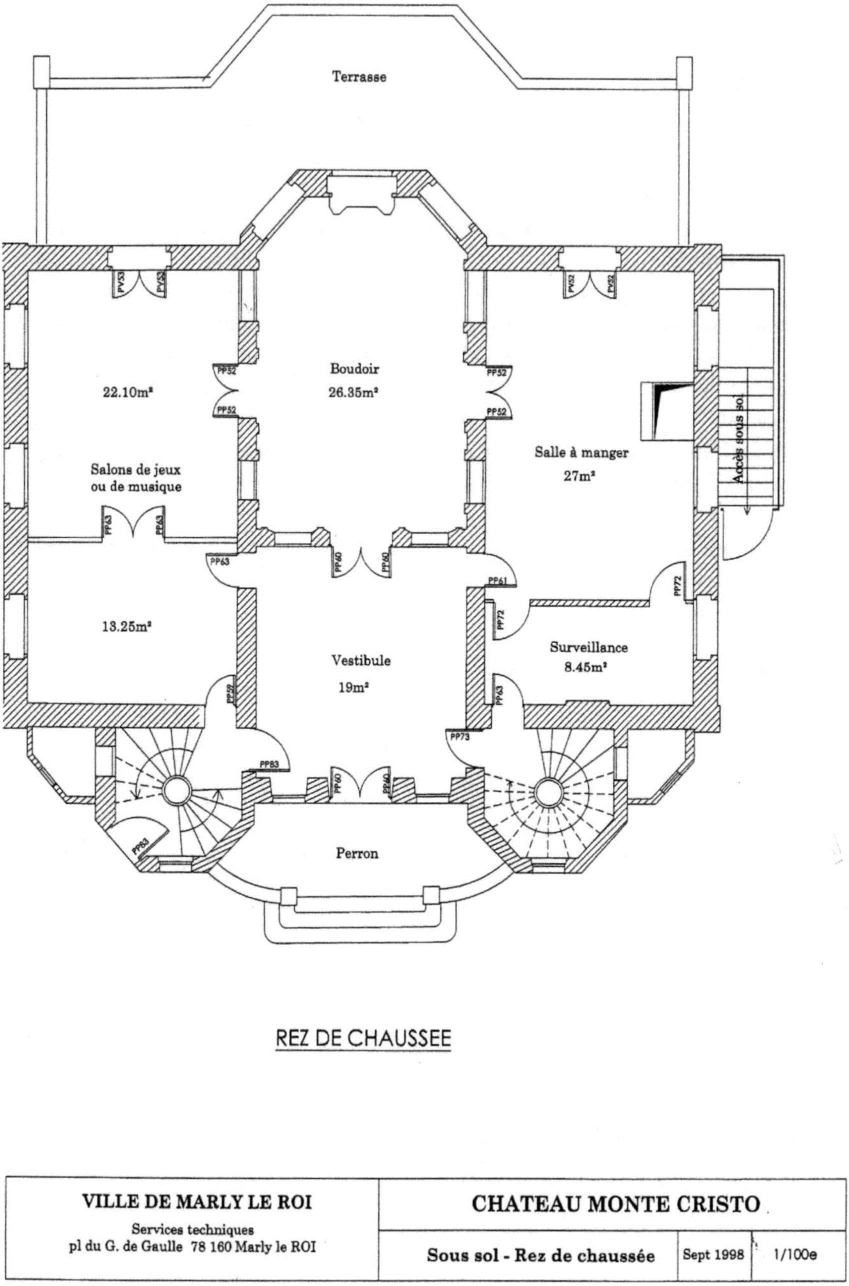

Fig. 4.12 Ground floor plan. Drawn by Marly le roi technic services

4.4 The Hexagon

4.4.1 The Hexagon and the Square

See Fig. 4.13.

The hexagon is the purist of the polygonal shapes because it has 6 equilateral triangles. It therefore has a natural organic intensity of cellular reproduction as witnessed with the beehive's structure. In Keith Critchlow's book 'Order in Space'[5], the growth of polygons is shown in a generational diagram with the equilateral triangle as the prime polygon followed by the square, the pentagon, the hexagon, the octagon, the decagon and finally the dodecagon. All of these shapes fit into a circle with each node touching the circle, but only the octagon and the dodecagon fits into a square because its main vertical and horizontal axis are equal in length and are at 90° to each other. This is also sympathetic to any rectangular component. Figure 4.14 shows some influential grid lines that would help in planning.

The following examples are beyond the norm in that one is experimental, and the other is for the wealthy, and therefor only serve as being useful from the view of the geometric planning.

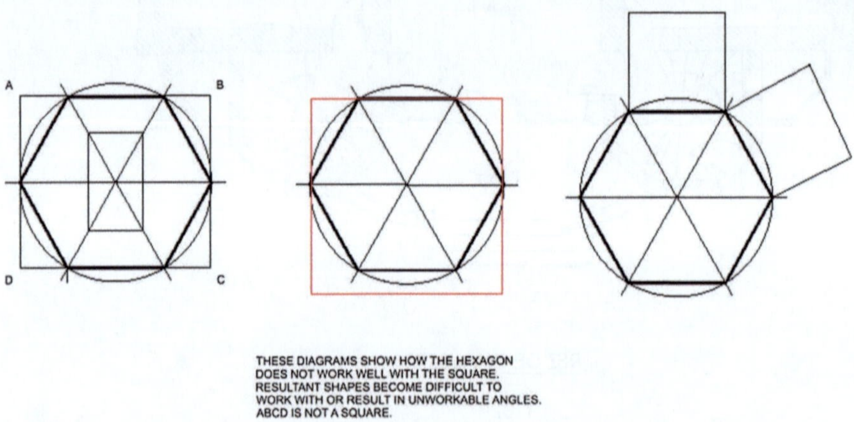

Fig. 4.13 Showing how the square does not relate well to the hexagon

[5]Keith Critchlow, *Order in Space*, Hong Kong, Thames and Hudson, 1968.

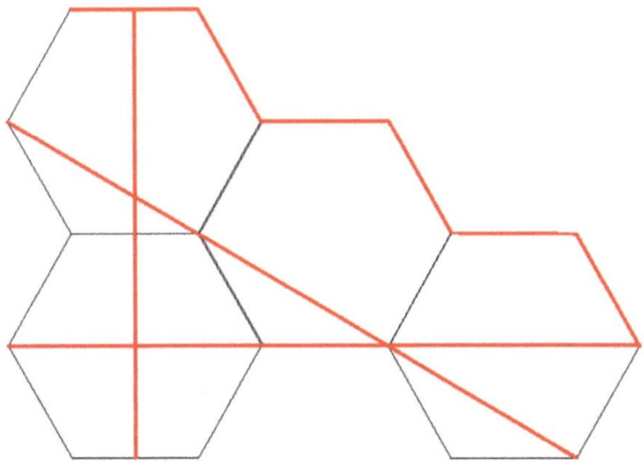

Fig. 4.14 Hexagonal grid lines

4.4.2 Dymaxion House

Buckminster Fuller[6] designed the Dymaxion House in the late 1920s but it was not built until 1945. It was Fuller's solution to the need for a mass-produced, affordable, easily transportable and environmentally efficient house. They were originally intended to be five-person family homes, with hexagonal floor plans and a central supporting mast. The mast was anchored to the ground by its base and offered the option of having an elevator installed within its frame. It provided 1600 ft^2 of liveable floor space. All the versions produced between this time were constructed of durable materials—mainly aluminium.

The plan in Fig. 4.15 shows two bedrooms with washing facilities, a utility room with kitchen facilities, a library and a living room. Unfortunately, the prototype did not lead towards any future manufacturing development. The restriction of the hexagon is that in its triangular segments, it produces a 60° internal angle such as in the utility room which is unworkable.

[6]Richard Buckminster "Bucky" Fuller (1895–1983) was an, author, designer and inventor.

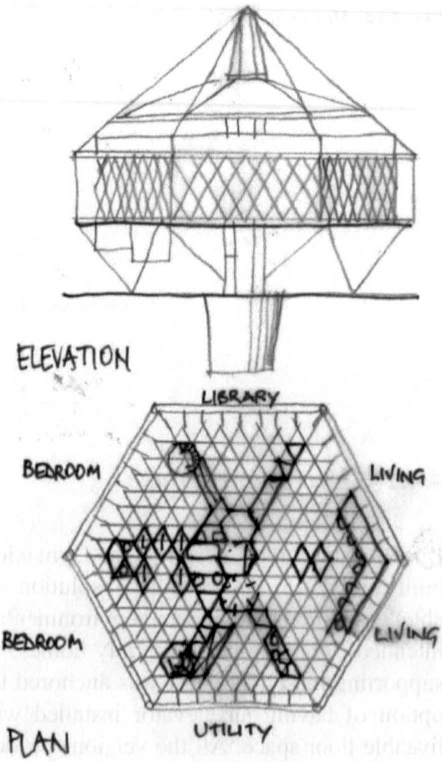

Fig. 4.15 Dymaxion house plan and elevation by Buckminster Fuller. *Drawn* Author

4.4.3 Hanna House by Frank Lloyd Wright

This bespoke house, nicknamed "Honeycomb House", is based on hexagonal geometry, with no right angles in the floor plan. This radical experimentation allowed Wright to explore open spatial planning. The house was designed in 1936 for Stanford University professor Paul Hanna, his wife Jean, and their five children. They lived there until 1975.

The exterior has all the trademarks of Wright's Taliesan Prairie style of design, yet it is not so clear in communicating the hexagonal planning. This is of course a dilemma common with much of architecture in that the geometry of the plan can sometimes be detached from the general 3D geometric form (Fig. 4.16).

The interior in Fig. 4.17 shows the hexagon at work with undulating glazing, floor pattern and the fireplace. An FL Wright approach was to carry through his geometric concept right through to the decoration as can be seen here with hexagonal cushions. The furniture has been designed to grow out from the hexagon in 3D but not as strict polyhedral forms. The hexagon cannot be extrapolated into a polyhedral form as can the pentagon, which makes the octahedron with 12 pentagonal faces.

Fig. 4.16 Hanna House, Stanford University Campus, California. 1936. Designed by Frank Lloyd Wright. *Courtesy* Stanford University

Fig. 4.17 Interior of the lounge with fireplace in foreground

Let us examine the plan: the hexagonal grid will follow various choices to suit the object in hand. You can see from the plan in Fig. 4.18 how controlling the hexagon has been with variations of extending walls on the grid. Great imagination has been used by Wright in determining what size a particular form or space is to be by modulating the basic hexagon. Wright has used the grid not to plan a repetitive

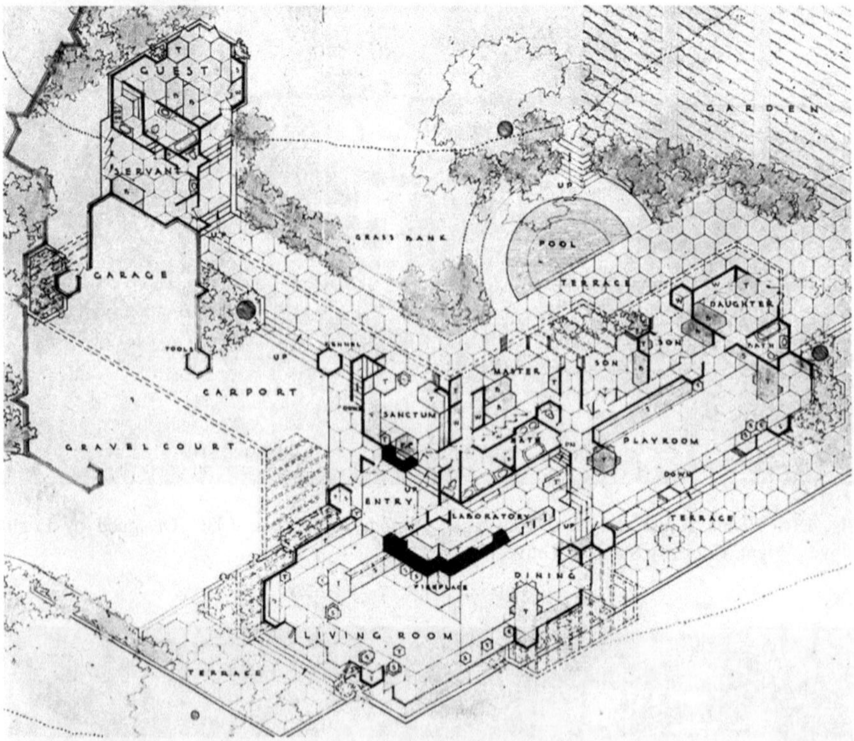

Fig. 4.18 Plan of Hanna house

cluster of cells like a beehive, but rather explode a variety of forms with space and structural linking devices as well as furniture forms. On the following house plan in Fig. 4.19, I have selected some different shapes out of the grid that suggest a parallelogram or the triangle for example.

Technical

Although Wright uses the term "wooden house" to refer to this house, it was built using a combination of red brick and wood, both inside and outside. The porch has San Jose, red-brick flat, as in the stack that forms the core of the house, with horizontal mortar joints of about one inch, while the vertical joints are flush. The exterior of the house is a combination of brick and wood. The outdoor deck is concrete with raised hexagonal modules to house plants and protect the old trees from the property. Large tables were secured with screws to the studs and the structural parameters were defined by a sequence of tables placed together horizontally. Using timber, stud walls allow the flexibility to change the space. Cypress wood was used for the horizontal slabs and hardwood for the frames.

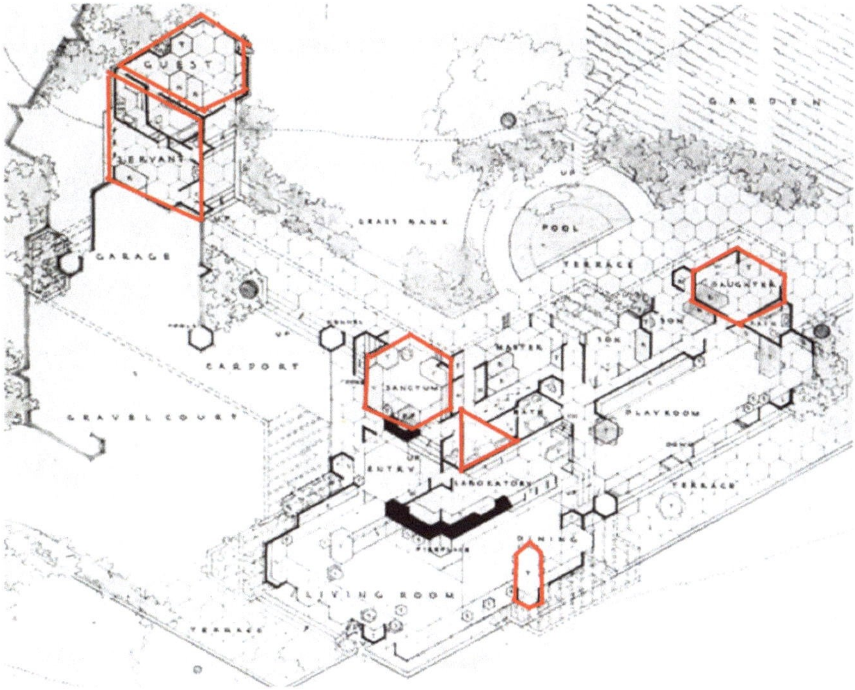

Fig. 4.19 House plan with identified variety of modulated shapes

4.5 The Octagon Development

4.5.1 Maths Block Parmiter School

It is very rare to come across any building based upon the octagon or of its parts, so I was so surprised to come across this beautiful two-storey Maths block of Parmiters School in Hertfordshire designed by architect David Lloyd in 2011. The use of the octagonal plan meant that the required area seemed larger length × width than if it was a rectangular space. Where the chamfered corners meet between the four room cells creates a circulation/service cross leading towards a central octagonal space with rooflight over (Figs. 4.20, and 4.21).

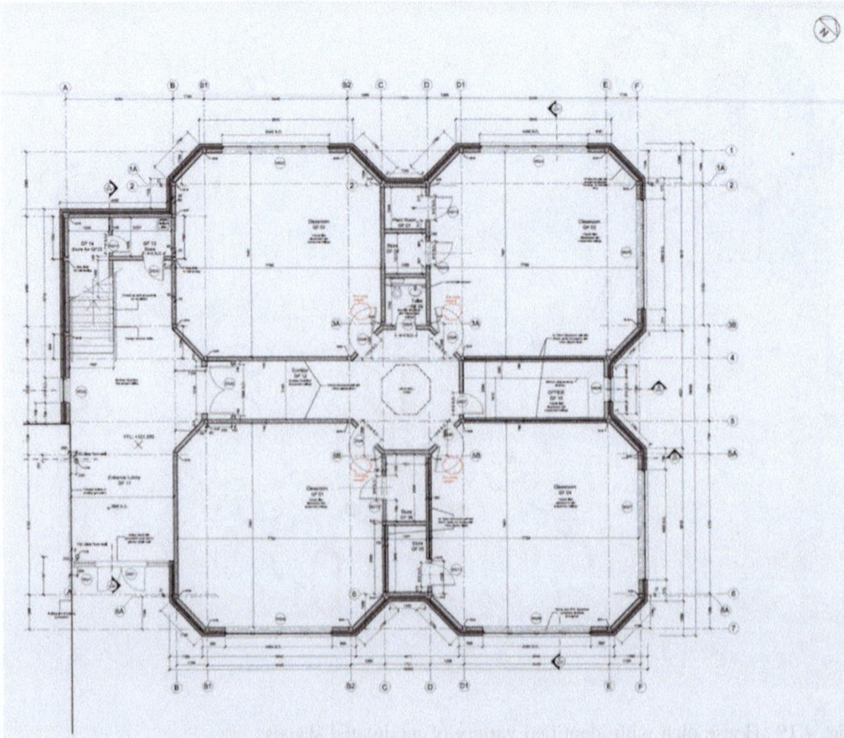

Fig. 4.20 Ground floor plan. *Courtesy* David Lloyd architect

Fig. 4.21 Exterior of Maths Block courtesy David Lloyd

4.5.2 Analysis of the Octagon

In Figure 4.22, left-hand diagram shows the nodes on a north–south axis as well as a possible subdivision using the nodal diagonals and a repeat smaller octagon, whilst indicating useable and unusable areas. The right-hand diagram shows the

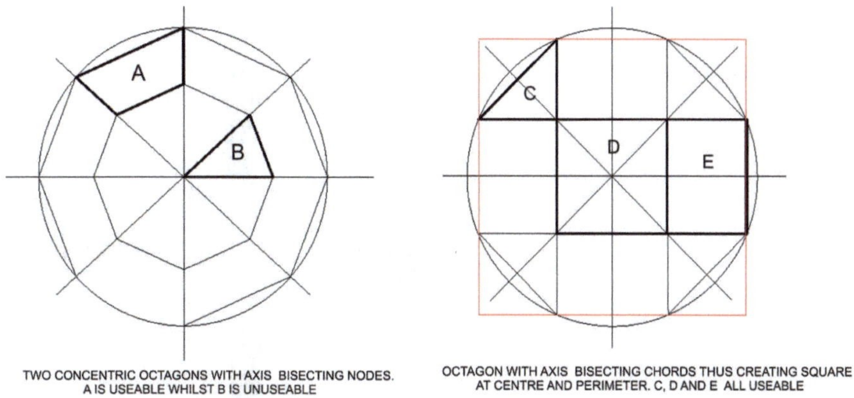

Fig. 4.22 Analysis of the octagon

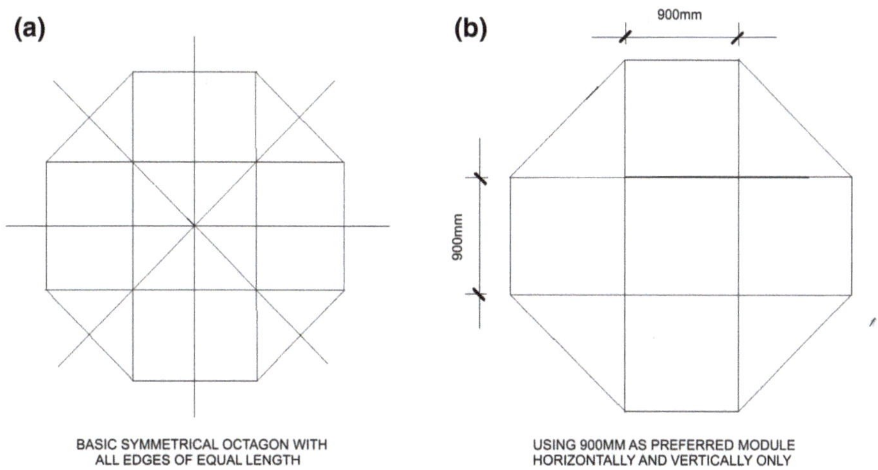

Fig. 4.23 **a** Regular octagon with congruent sides and angles, and **b** an irregular octagon with the 900-mm chosen module of the square grid but NOT the angled sides

sides centralised on a north–south axis and a repeat of the square indicating better useable areas

You can now see how the diagram in Fig. 4.4 fits neatly into diagram (b) in Fig. 4.24 and the resulting expansion shown in diagram b facilitates circulation access. Diagram b shows in grey possible useable areas. This is an important generating cellular shape that would suggest the angles planes could be windows or door access, but could also provide an emotional respite from boxy rooms. This is the planning orientation that I shall develop in this book. I have decided to explore the octagon as a basis for cell development and for the planning of a house unit. Figure 4.25 shows how cells C and D could interconnect and Fig. 4.26 shows a popular use of the irregular octagonal shape.

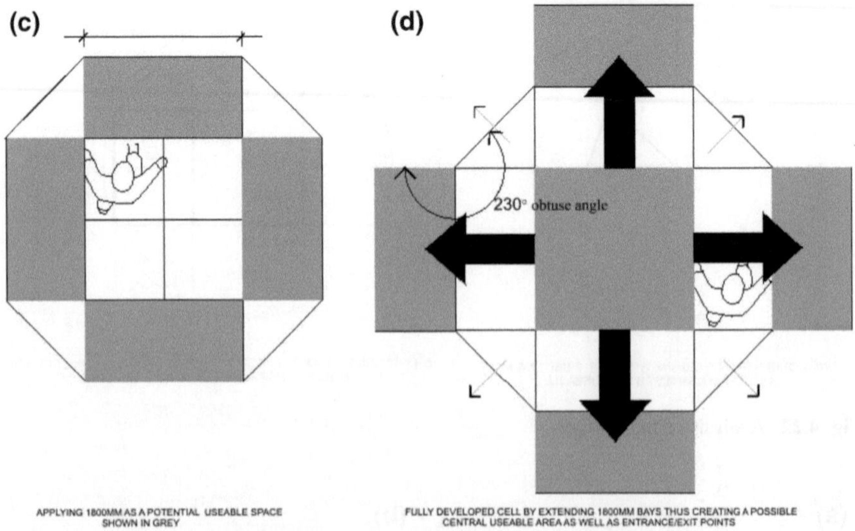

Fig. 4.24 **c** Compact useable space and areas designated by the double module width 1800 mm. **d** Expanded version to allow for entry points and 8 × 230° obtuse angles

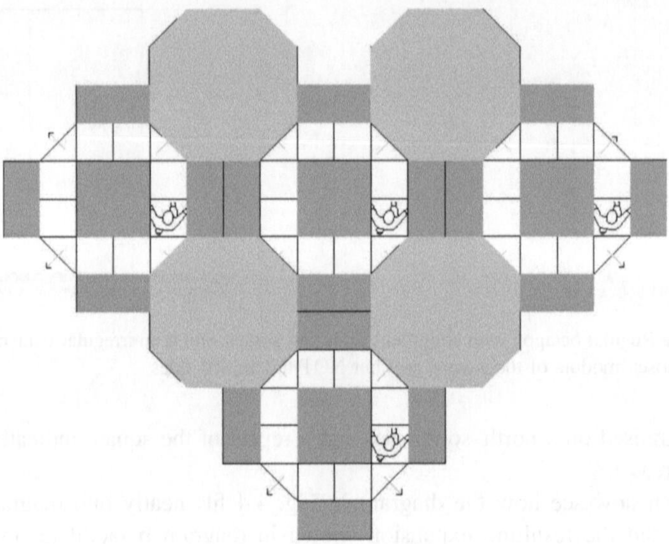

Fig. 4.25 Showing **c** and **d** from Fig. 4.24 but interlocking to create a pattern repeat

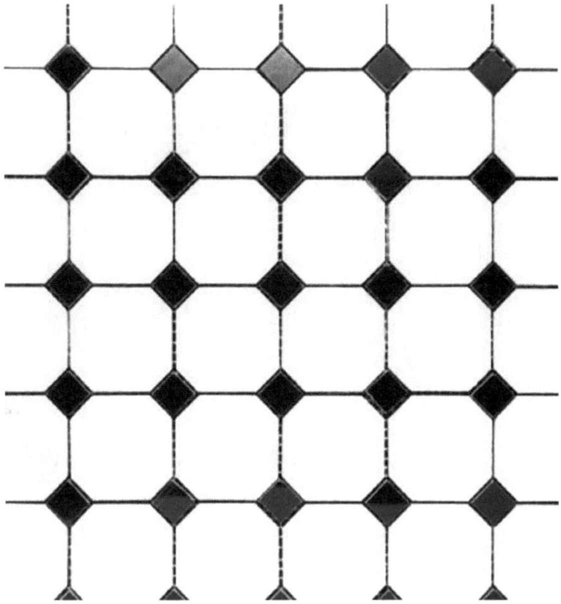

Fig. 4.26 A popular irregular octagonal ceramic tile layout with small squares

4.5.3 The Polygonal Effect

We must release ourselves from the overriding and suffocating discipline of the square ; the box which stakes its economic claim for being easily reproducible and makeable. The polygonal shape is organic and friendlier and deserves our attention (Fig. 4.27).

The bay window is so attractive to people of all ages, but seemingly now it is a privilege. I grew up in the top floor of the three-bedroomed Victorian flat in the middle photograph (Fig. 4.28).

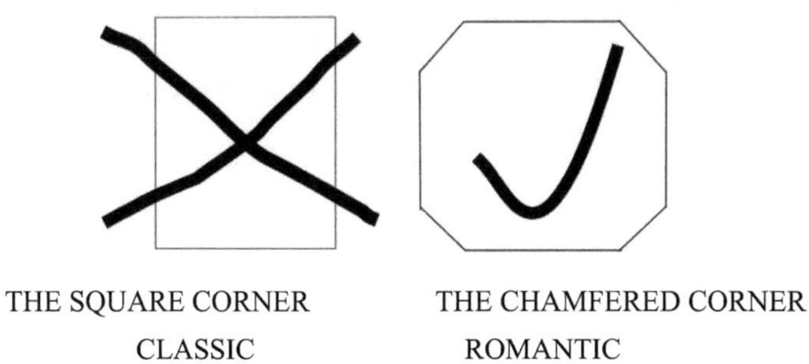

Fig. 4.27 Diagram showing the square corner and the chamfered corner

Fig. 4.28 Bay windows of period buildings in the UK. *Photographs* Author

The polygonal shapes offer far more interesting possibilities than the severity of the square grid. The variety of directional choices, angles of potential spatial divisions and glazing, all contribute to a far more welcoming and comforting interior. The following diagrams Fig. 4.29a, b, c explain how the straight single line in geometry grows by accompaniment of other straight lines into a square corner, which then becomes a square—a dead end. It then shows that if the second line is at an angle of 45°, it can be joined by a third line creating a far more restful situation that can repeat itself less dogmatically that the confines of a 'box'. A box is a prison; it is for putting things in. So, an octagon is suggestive, but a square suggests nothing. The square has four corners and sides. Four resonates with practicality, organisation, loyalty, mastery, conservatism, solidity—whereas the octagon has eight sides and eight corners. Eight resonates with infinity, hope and everything that is good in the world. Its called 'wellbeing'.

Figure 4.30 shows the interior of the octagonal garden room in the grounds of the sixteenth century Melford Hall in Suffolk, UK. You can see how more relaxing the atmosphere is with walls and windows at 45° to each other as opposed to sitting in a box.

The window in diagram Fig. 4.31 below shows how the wall surround appears dark in contrast to the daylight coming through the window. Figure 4.32 shows how a chamfered wall reflects the daylight and hence makes the room brighter.

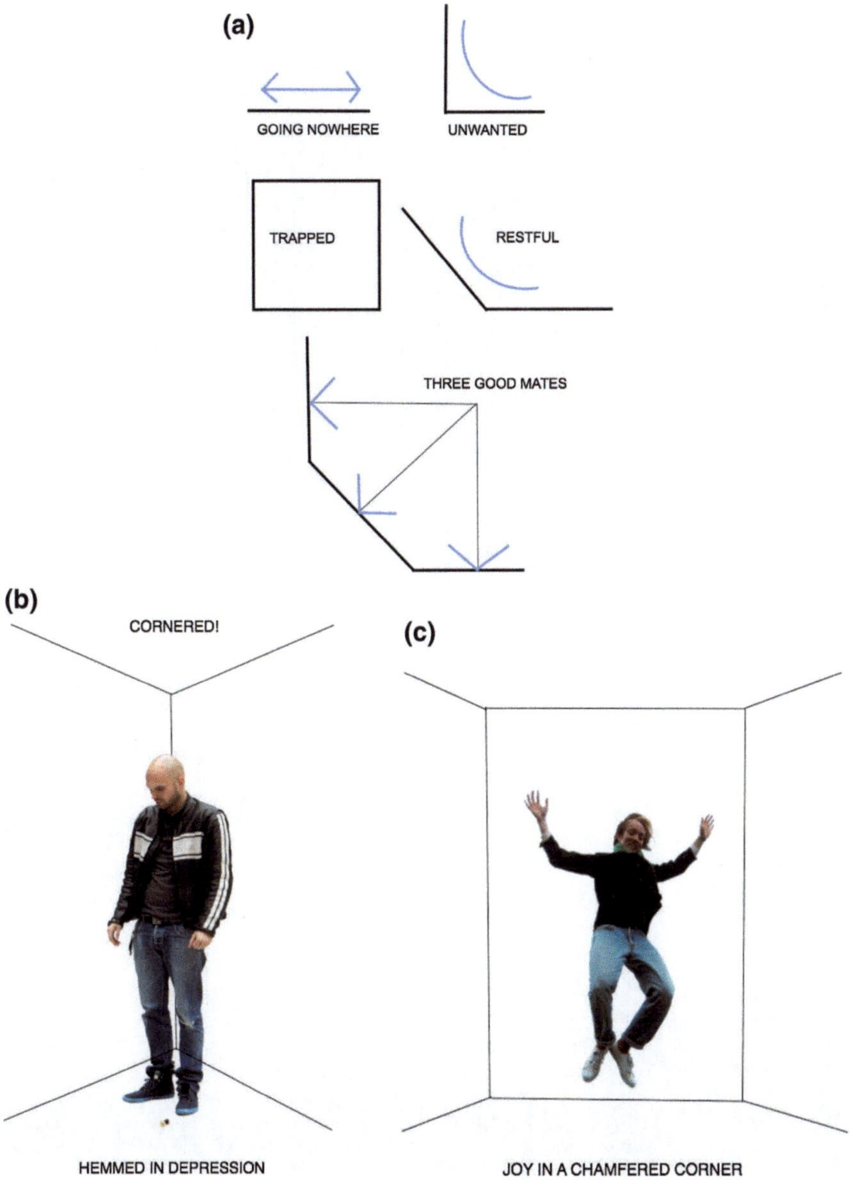

Fig. 4.29 **a** Geometric progression of line to part polygon, **b** 90° corner depression, **c** chamfered corner elation

Fig. 4.30 Interior of garden room. Melford Hall, Suffolk, UK. *Photograph* Author

Fig. 4.31 Square room with dark window wall

Fig. 4.32 Room with chamfered walls reflecting window daylight

4.5.4 Wellbeing

The effect that interior space can have on people is borne out by the following recommendations:

> *Having somewhere safe and warm (literally and figuratively) to live is fundamental to our mental health and wellbeing. Housing should provide not only shelter but also a secure and positive environment that supports people as their lives progress.* Mental Health Foundation 2016

Adequate and appropriate housing is widely acknowledged to be a crucial underpinning of health and well-being. National Mental Health Development Unit

The Design Research Society held a 50th anniversary conference in Brighton 2016 and the subject matter seemed to me appealing, but after wading through the published document of these proceedings I finally came across a paper that I thought would help me with this book entitled 'Exploring Design for Happiness in the Home and Implications for Future Domestic Living'. I am afraid that it focussed on behavioural matters with objects and the psychology of personal relationships as opposed to the 'design' of the spaces. It did however confirm the following:

Our current way of life is unsustainable. Current GDPbased economies affect the environment, economy and society; they are reliant on high levels of material consumption, productivity, create excessive waste, and motivate neurosis. These lifestyles are not making us happier (Hofstetter et al. 2006; Diener and Seligman 2004) and have been linked to higher levels of stress, anxiety and depression (Kasser et al. 2014; ibid).

Approximately 450 million people worldwide have mental health issues and 1 in 6 people are affected in the UK every year (WHO 2001).

The RIBA published a report in June 2017 which highlighted the following:

It warned the lack of space and light could cause marital problems and even stop children from reaching their full potential at school.

'A lack of space has been shown to impact on the basic lifestyle needs that many people take for granted, such as having enough space to store possessions or even to entertain friends.In more extreme cases, lack of adequate space for a household has also been shown to have significant impacts on health, educational attainment and family relationships.'

This concern with space is the crux of my argument in Chap. 5 regarding my own design proposals. There is a proliferation of research material that deals with sustainability and the socio-economic status and well-being of people, but not enough realistic analysis on what kind of spaces do people feel comfortable in. This is why I have explored the octagon to help with providing that welcoming element in the interior. The following quotation also confirms the importance of well-being:

Contrary to popular belief, many of the luxuries commonly associated with high end residential developments are no longer limited to the affluent alone. A growing consciousness of all things beneficial to health and happiness is resulting in a trickling down effect and a widespread societal demand for healthy living environments. Today's residents want convenient shared amenities which encourage a sense of community and promote physical and psychological wellbeing. Ben van Berkel, Founder, UNStudio. November 2016 at World Architecture Festival's 'Housing for everyone

A high level of visual information is not a matter of 'beauty'. It helps to establish the individual character and atmosphere of a place as well as our psychological involvement. Both aesthetic and symbolic information are necessary conditions for the identification of the environment, which is endangered by the growing uniformity and anonymity of our cities.

Theo Crosby.[7] 'How to play the environment game'.

This quote is from the book, which was the record of an exhibition held at the Hayward Gallery in London, which I went to see in 1972. This emphasises how the visual qualities of an environment help to consolidate the whole experience to people and that the identity of place needs to have that re-assurance, in this case for our occupiers of housing.

4.6 Other Historical Octagonal Examples

Although these examples have nothing to do with my housing quest, it is the use of the octagon in plan form that is of interest, by indicating how powerful the form is as an object and as an enclosing form.

4.6.1 Plans—Basilica of San Vitali, Tower of Winds, Dome of Rock

Please refer to Biltmore House in Chap. 1 Para 1.3.2 and see the number of octagonal or part octagonal room shapes there are in the plan. Early domed buildings such as the Basilica of San Vitale in Ravenna (plan in Fig. 4.33) and Florence Cathedral used the octagon in plan form as a structural springboard for the dome itself. One of the earliest known octagonal structures is the 'Tower of Winds' in Rome built about 50 AD (plan in Fig. 4.34).

The Dome of the Rock in Jerusalem is an example of a religious building with an octagonal ground plan. The construction began in 684 AD and rebuilding took place in the eleventh century. The dome was placed over *as-Sakhra* (the rock), on the (supposed) geographical position where the prophet Mohammed ended his Night Journey to Jerusalem and ascended to heaven. The addition of the gold-plated roof took place in 1959–1961 (Figs. 4.35 and 4.36).

The 'evolution' of the ground plan from a circular outlay (I) to the cross (II) and the square (III) is best seen in the octagonal and polygonal features of the ground plan (IV). Subsequently, the (quadralectic) circle is round again when the polygonal scheme merges with the circle, in a new first quadrant (Fig. 4.37).

[7] *How to play the environment game.* Theo Crosby. Harmondsworth. Penguin Books Ltd. 1973.

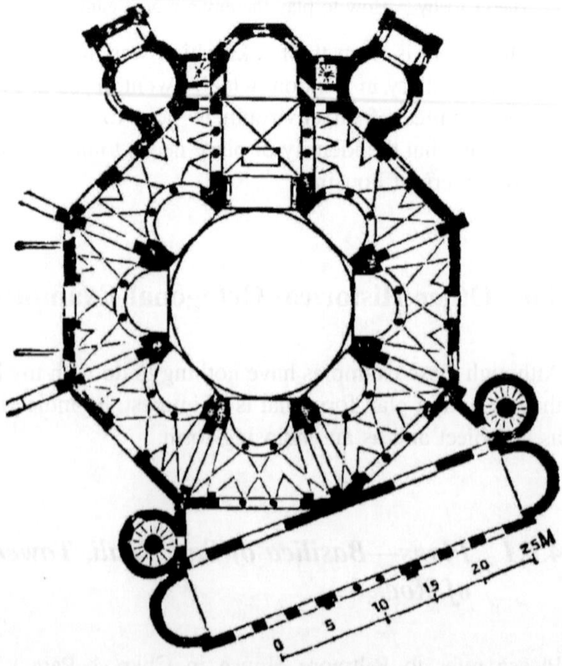

Fig. 4.33 Basilica of San Vitale in Ravenna, Italy 526 AD. Wikipedia, author Byggnadskonsten, San Vitale i Ravenna, Nordisk familjebok.png

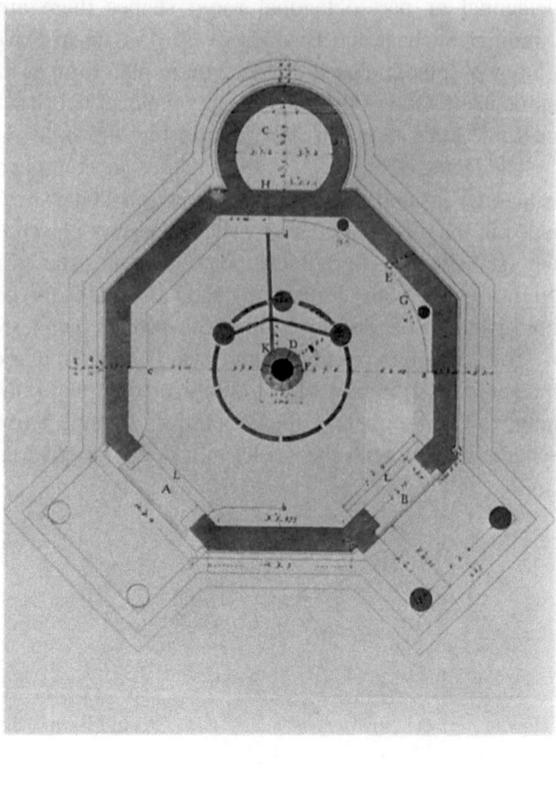

Fig. 4.34 'Tower of Winds' Horologion of Andronikos Kyrrhestes is an octagonal Pentelic marble clocktower in the Roman Agora in Athens. Wikipedia

Fig. 4.35 Dome of the Rock on the Temple Mount, Jerusalem, 684 AD. Photograph from Wikipedia Commons, *Source* Zairon. Marten Kuilman

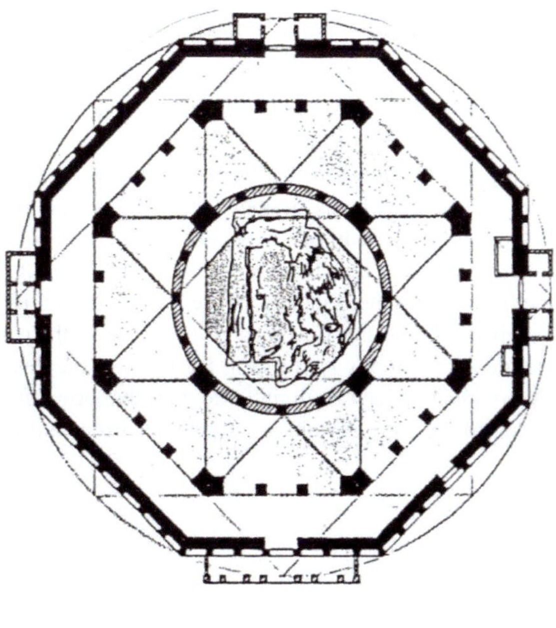

Fig. 4.36 Plan of Dome of the Rock on the Temple Mount, Jerusalem, 684 AD. Wikipedia Commons, *Source* Zairon. Plan from Marten Kuilman

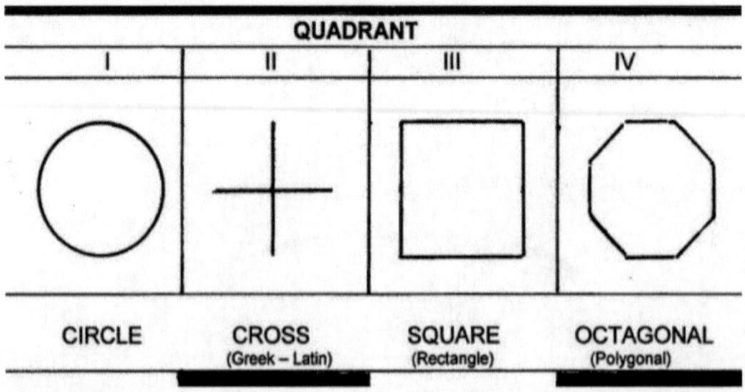

Fig. 4.37 Geometric development of the octagonal plan. *Source* Marten Kuilman, Quadralectic Architecture. 2013

4.7 American Octagonal Examples

According to James Wentling[8], America did not adopt a national style of architecture as the progress of housing depended upon the importation of European and what is often referred to as the 'Colonial' style.

4.7.1 Palliser's New Cottage Homes

These were built according to inherited styles mainly from Europe, which is what the Victorian period was all about. George Palliser (1849–1903), an English immigrant, first worked as a carpenter in Connecticut, then owned a sash, blind, and door factory. With his brother Charles, his publishing business produced eleven titles, in multiple editions over 30 years. These pattern books contained plans and guides on how to build a variety of designs. Although built of standard construction methods, many thousands of houses were built all over the country. The house that caught my eye is this one because of its use of the octagon. Four octagons around a central square lobby space form the basis of the design spoilt by inconsistent other square bits that have nothing to do with that powerful geometry. The square corners added to the dining room are especially irrelevant (Figs. 4.38 and 4.39).

[8]James Wentling, *Designing a Place called Home*. Switzerland, Springer. 2017. P2.

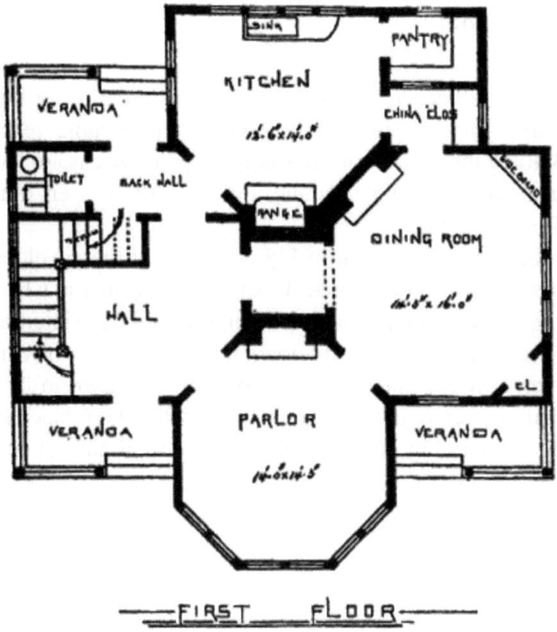

Fig. 4.38 Plan of a typical Palliser Victorian House. Downloaded from Internet archive

Fig. 4.39 Exterior view of a typical Palliser Victorian House. Downloaded from Internet archive

4.7.2 Topsider Homes

In America, there have been many other houses designed using the octagon as a planning basis since the mid-nineteenth century and a more recent example is shown in Fig. 4.40 by the American Topsider company. Areas are lounge including desk, kitchen including washing machine, dining, hall, 1 × en-suite double bedroom, two double bedrooms.

The main problem with this plan in Fig. 4.40 is that the envelope may well be octagonal in shape (as with the historical examples) but that is where its influence ends as there is no attempt to replicate it with any related form internally. The centre is acknowledged as is the vertical and horizontal axis for partitions and furnishings. The resultant internal room shapes are not specially designed to be as they are.

For example, if you remove the octagonal perimeter and replace it with a square as I have partially done in Fig. 4.41, this shows how weak the octagonal concept is.

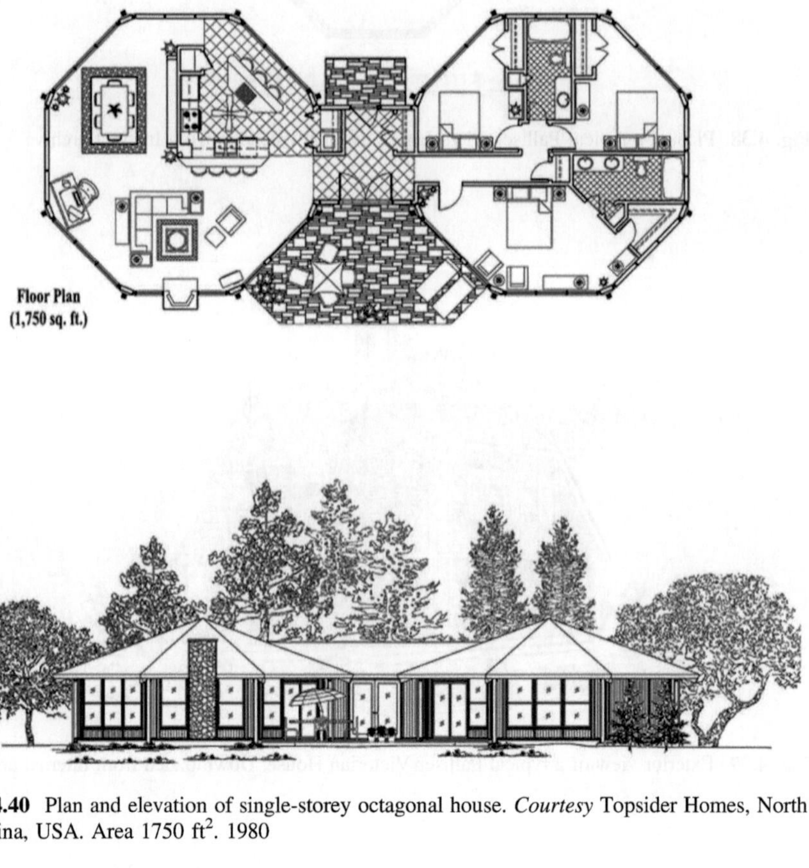

Fig. 4.40 Plan and elevation of single-storey octagonal house. *Courtesy* Topsider Homes, North Carolina, USA. Area 1750 ft^2. 1980

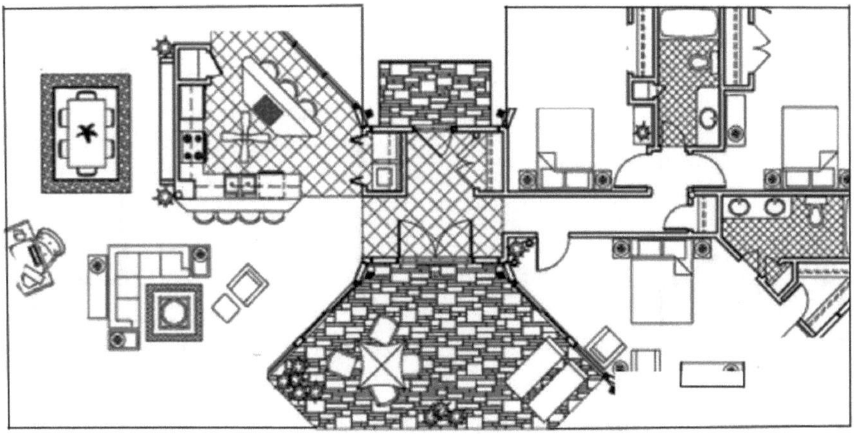

Fig. 4.41 Topsider house plan adapted by removing the octagonal perimeter

4.7.3 Hoge Home

I am showing the plan below, which is of a luxury detached house, because of the octagonal shaped rooms and the resulting angular spaces that result. It is not an octagonal house. It especially works well in the foyer with a symmetrical angular peeling off to either right or left, which is so much more inviting than 90° corridors. As with houses of this size in America, there is plenty of space, which is not really necessary. It so happens that the majority of suburban homes in the USA are detached amounting to 90%, and of course, they have the space to have such expectations. Homebuyer polls suggest that this is part of the American dream. Today many suburban communities consist of excessively wide streets and oversized housing plots, with the garage asserting itself visually to such an extent that the house behind becomes subordinate architecturally. Increasingly, the interiors of new houses have become the main selling points of the market for prospective buyers. But this concentration is carried out on a mainly decorative basis and not in any revolutionary sense with regard to planning. Sure, America lead the way with more 'open-planned' ground floor spaces but otherwise the houses were based upon strong traditional and conservative values. Sadly, many houses are not designed by architects but are built from pattern books used by the developers (Fig. 4.42).

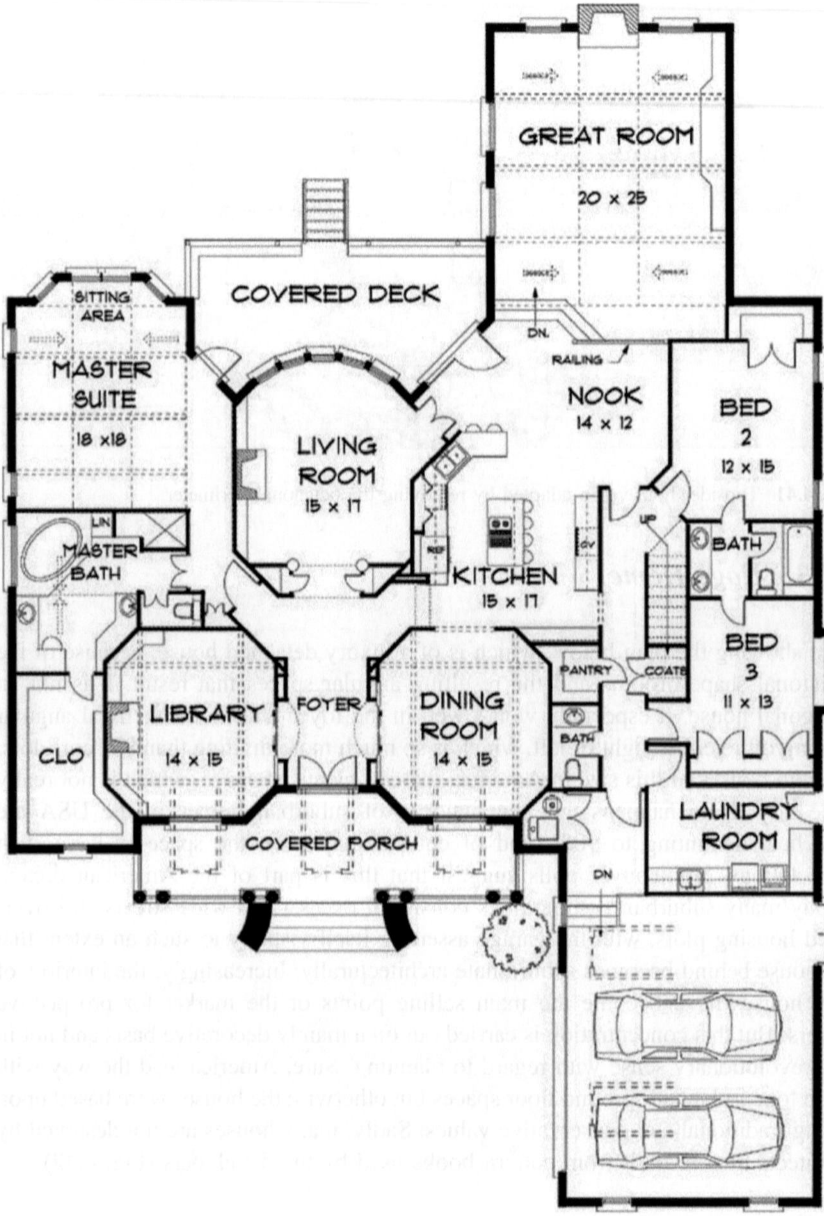

Fig. 4.42 Plan of three-bedroom house by Rick Hoge of the Hoge Group, USA. 2016

4.8 Conclusion

I am writing a protest book and hopefully explaining the true situation that exists in the housing industry that may draw some sympathy from readers, albeit there will be people who may argue against my thesis. But it is no use protesting unless I have something positive to say about a way forward. This chapter prepares a formalisation of a geometric approach that encapsulates the spirit of living within a sympathetic architectural discipline.

Fig. 4.43 Student project using the dodecahedron. 1978. *Photographs* Author

Fig. 4.44 Student project using the dodecahedron. 1978. *Photographs* Author

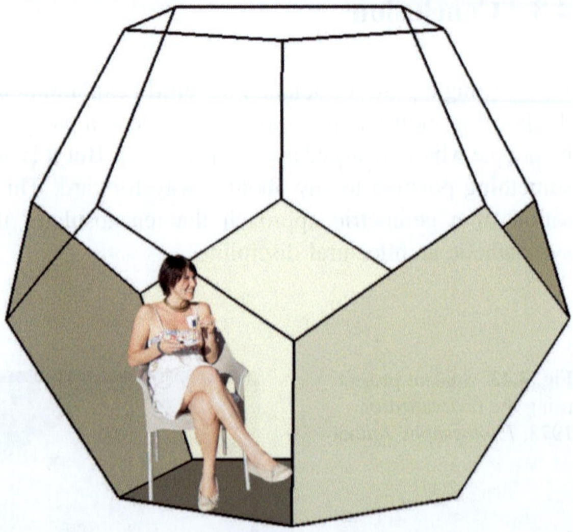

Fig. 4.45 Frame dodecahedron showing the partial panels as above. *Drawn* Author

I ran an experimental student project some years ago whereby part of a dodecahedron was made of expanded polystyrene with plywood fixing nodes as Figs. 4.43 and 4.44 show. To build a mock-up, full size was invigorating and inspirational (Fig. 4.45).

Chapter 5
A Project Proposal in Two Stages

Innovation doesn't come from the profit motive.
Innovation comes from those who are happy to embark on a course of action without quite knowing where it will lead, without doing a feasibility study, without fear of failure or too much hope of reward.
Frank Cottrell Boyce—Guardian 16.7.16 Writer.

Abstract This chapter is devoted to my own idea of designing an estate house in concept form only and at a later stage to be prefabricated for mass production whilst using sustainable methods. It is an attempt to address the issues outlined in previous chapters assisted by helpful consultative expertise. I am presenting Stage 1 as a cellular approach for the planning of activities, and Stage 2 a more realistic plan derived from Stage 1 but as a more compact solution. The final desired scheme is reduced even more in floor area to be more competitive with the marketplace. I propose three-roof design solutions to the same plan. The whole series represents the design process at work. The design approach for anything that is produced in a factory must be similar to designing a product such as a kettle or motorbike. Le Corbusier (Le Corbusier. 1887–1965. Charles-Édouard Jeanneret, known as Le Corbusier, was a Swiss-French architect, designer, painter, urban planner, writer, and one of the pioneers of what is now called modern architecture. He was born in Switzerland and became a French citizen in 1930.) wrote "A house is a machine for living in," in his 1927 manifesto Vers Une Architecture (Towards A New Architecture). This was not to be taken literally but was rather to inspire an approach embracing efficiency of control over aesthetics and function. The product designer considers all the 'working parts' and decides through customer needs and research what is seen or expressed and what is contained or 'hidden'. A house exposes rather more proportionately to an engineered product both inside and outside. There are degrees of containment in a house from common usage spaces to more private spaces; from primary storage (open shelving to kitchen/utility cupboards, wardrobes) to secondary storage (generally attic spaces). I am well aware that architects, developers and local authorities have to work within cost constraints to make a project financially viable and they also have to ensure that it falls in line with current building regulations. It is my sincere belief that current regulatory

A. Sully, *The Estate House Re-designed*,
https://doi.org/10.1007/978-3-319-90397-2_5

space standards need an overhaul as they do not serve the public well. I am also aware that since the concept of the tree-lined suburban avenues of the past idyllic suburbia, the population whom we serve have not been given appropriate living accommodation that gives them the human dignity that they deserve. As you have seen in the Case Studies of Chap. 2, there are some beautifully well-designed complexes that have won awards for their improved contribution towards providing a housing community. The main problem is that architects seem to be caught in the trap of boxy modernism that denies any geometric variation when it comes to the interiors. Once pig-pen architecture along with its cost constraints takes control by being given the green light from the authorities and stakeholders, it is a strong controlling factor. Realising that I have an uphill task, my proposals are in two concept stages only, and as with any design project, it requires the makers to make their contributions according to the cost constraints imposed. I totally accept that the detailed construction of the structure and services will affect the finished article but I am also aware that many variations are possible. In my proposals, furniture is indicative as occupants will obviously personalise their own spaces. My position can also be explained by the following quote by Jamer Hunt in the book 'Design Anthropology' (Alison J Clarke editor. *Design Anthropology.* Chap. 6 by Jamer Hunt 'Prototyping the Social: Temporality and Speculative Futures at the Intersection of design and Culture.):

> *There is a fringe of design practice that does adopt a more ideologically driven position in relation to design's place in the world, and that work begins to point to a fruitful conjoining of critical social analysis and design. These projects envision design futures through the technique of scenarios. They stretch our current conditions to their logical extreme, extrapolating the present into the future. Like science fiction, these scenarios of a designed future foreground the politics of our current manias. In that way, they implicitly challenge our designed—and social and political orthodoxies. The projects effectively aim to give form to emergent social practices, providing us with a distorting mirror through which to see our present. The scenarios, by and large, are social scenarios in which the material and immaterial built environment prompt new kinds of behaviours.*

5.1 Cellular Approach

5.1.1 Cellular Thinking

The following diagram in Fig. 5.1 explains how my cellular approach varies from the traditional approach with regard to the containment of domestic activities which I shall just label A, B and C for example. The cell under 'My Approach' wraps around activity A with a closeness of provision, whereas the 'Traditional' house

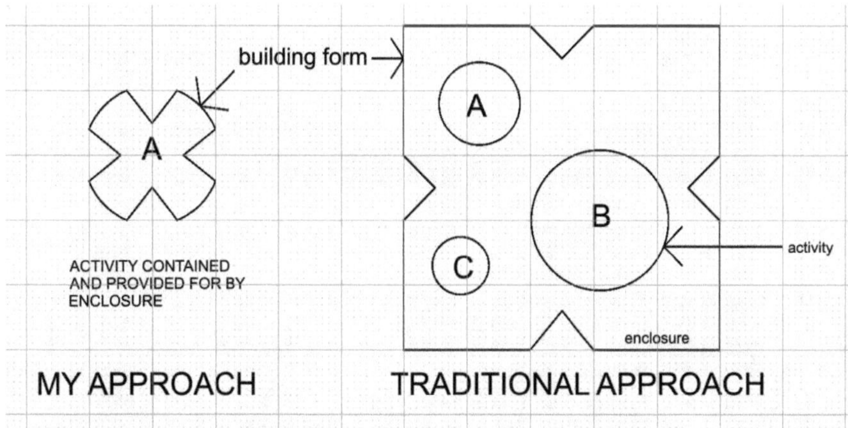

Fig. 5.1 Cellular approach diagram

plan's building form with activities A, B and C shows each activity detached from the enclosure, and hence, the walls, floor and ceiling are considered separately from the furnishing in many cases. Traditionally, walls are positioned to follow area requirements rather than activity needs and windows are positioned mainly with regard to the exterior elevation appearance rather than the daylight/view needs of the interior.

This results in the argument that my approach could be described as being inflexible, whereas conventional open layouts allow for different arrangements. The problem with the notion of flexibility is that it works fine in sports halls for example, whereby they can be adapted easily for many different sports, but in a house with carefully specified activities and facilities it is essential for the peace of mind and well-being of the occupants that it works easily, with minimum of effort, and is a delight to live in. If we look at the living space through history, the high decoration of the interiors had a strong integration of enclosure to furniture. For example:

Containers
Walls, floor, ceiling, doors, windows, fireplaces, niches and built-in bookcases were designed to integrate with the architecture of the building.

Contents
Furniture such as bookcases, cabinets, chests, tables, mirrors, lighting, sideboards, piano, bureaus, fabric/upholstery would also be designed with decorative features similar to the architectural detailing of the building even though they were free-standing and moveable.

Activities (A, B and C)
Conversation, card/board games, refreshment of coffee or alcohol, embroidery, reading, writing, entertainment such as music and singing.

Fig. 5.2 Reception room of Hawkstone Hall, Cheshire, UK. *Courtesy* Weston Hall

The interior of this country house in Fig. 5.2, whilst far away from my focus in this book, shows various groups of furniture where people could take part in any of the above list of activities because they had the space to do it. The reason for the strong bonding of all of these is that the space is articulated and defined by frames and borders that give the contents a resting place. There are axis and focal points, which are almost completely absent in modern housing. Just because we are dealing with much smaller spaces does not mean that defined co-ordinates of structure and spatial definition need be absent.

Open-plan homes that appear to offer flexibility instead offer mixed messages of uncertainties with overlapping functions that contribute to an unsettling experience and a lack of purpose. A large space gives a false impression of high wealth status, which is a misnomer. To be analogous with clothing, it is rather like putting someone in a sack whereby they have freedom of movement, instead of inside a costume tailored to fit all limbs and torso of the body.

5.1.2 The Windows—With 900-mm Cill Height Wall (See Sect. 1.6.4)

The full-height windows that I am dismissing in my proposals also suffer from an unsightly array of net curtains and blinds which are more noticeable because of the larger area of glass compared with a smaller window. Below is a 3D view (Fig. 5.3) of

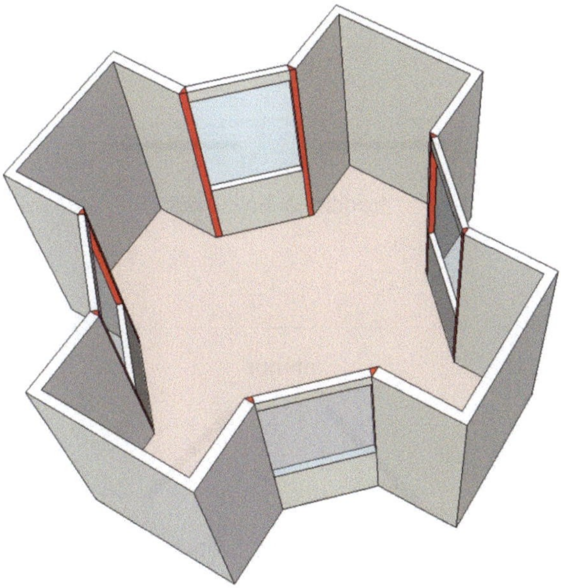

Fig. 5.3 3D view of larger cell

the larger cell from D in Fig. 4.24 with windows in a suggested position on the diagonals (inverted bay window) leaving the rectangular recesses to be the activity/ storage areas. The red indicates a joining function. Please note that in Stage 2, I propose swapping the windows for the rectangular recesses to test such validity. The important point to stress is that the windows have a place architecturally, whereas in traditional rectangular rooms they are positioned NOT necessarily in relation to the interior but according to how well they look with regard to the external elevations. Windows usually have a loose connection with the building form in terms of location rather like the furniture in Fig. 5.7. There are exceptions I know. The aim is to incorporate metal roller shutters over all doors and windows as a security measure.

The following drawing shows a comparison with my inverted bay window with a traditional bay window. The interesting point about this is that the traditional bay is about thrusting out beyond the wall inviting more views outwards, whereas the inverted bay is more reticent and modest in its relationship with the exterior. It is also more dynamically related to the containment structure and therefore has a more dutiful job in the interior than the traditional bay window (Fig. 5.4).

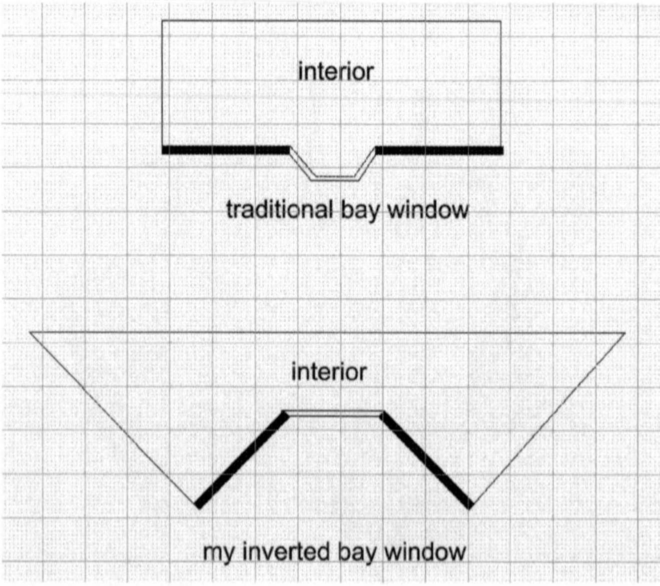

Fig. 5.4 Bay window comparison

5.1.3 Spatial Organisation

As I have said before, floor to ceiling glazing I consider to be wasteful of heat loss and psychologically undesirable with regard to security. Who wants to stare out at patio paving? Any 'views' are usually above the waist. A solid upstand below the window provides more satisfying containment and space for any furniture against it as well as more comforting for children playing against it. My planning versions also follow the wide house width with central entrance axis as opposed to the very common narrow width house with entrance to one side. The narrow width houses (inherited form Victorian terraced/semi-detached houses) have the unfortunate long access corridor, due to the straight staircase, which I consider unsafe and it has less garden privacy than a wide width house. The wide width house favours the more safe dogleg staircase as the diagrams in Fig. 5.5 (see case studies diagrams). It is universally accepted that a basic family house will be two storey as this compactness is more economically efficient for land occupation as well as building services and insulation. Various combinations of roof designs are explored between pitched and flat roof.

From the list of main activities in the last chapter and Fig. 3.3 from Chap. 3, and change of names from Sect. 3.2.3, the following diagram (Fig 5.6) shows how the major split has logically created two floors in order to achieve higher density use of the plot per m^2: Ground floor has the busiest activities (5) whilst the first floor contains the sleeping/washroom arrangements. The workspace is an activity that could be placed

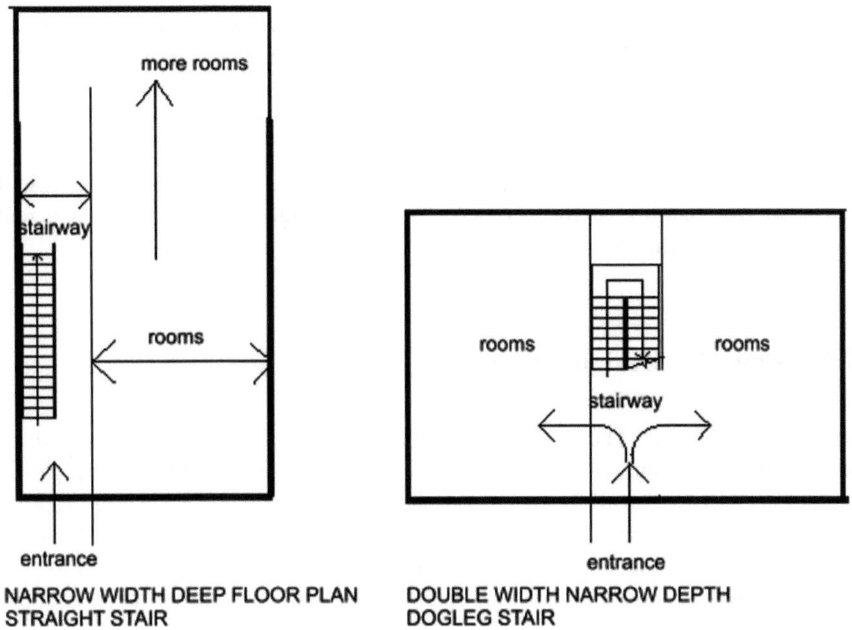

Fig. 5.5 Plan diagrams showing two common house widths

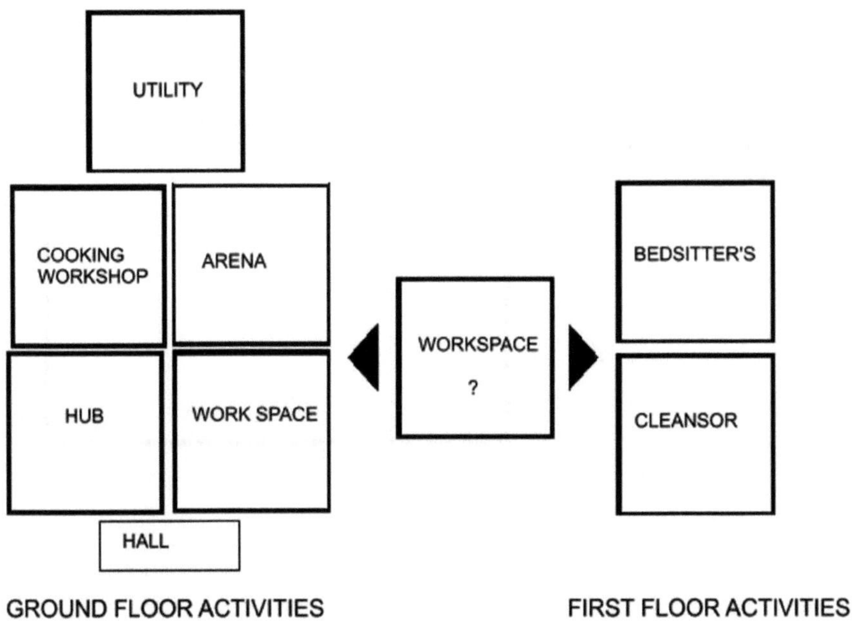

Fig. 5.6 Ground and first floor suggestions of activity placement per floor with new place names: cooking workshop (kitchen), arena (living room), hub (dining/playspace), workspace (study), bedsitter (bedroom), cleansor (bathroom)

on either floor but my preference is for it to be on the ground floor as the use of IT and media has become an all consuming regular pastime that suggests being on the noisier floor. It also operates with a diminishing use of books and paper as an accompaniment. Most domestic activities have their own priority time of use, but use of media devices has become an almost continuous activity.

The drawings of the following Stage 1 are presented more as diagrams because it is testing the viability of the geometry, whereas Stage 2 is presented as a real building project with 250-mm-thick external walls.

5.2 Stage 1—Design Proposal 1

Den is not included. The area of this plan is way over the size that could be competitive within the marketplace but planning is a sequence of beginning with a full pot in order to discard the extraneous. The emphasis is containment of furniture (as explained in Sect. 5.1.1) by spatial form as opposed to the traditional domestic plan of furniture either loosely positioned against a wall or in the middle of the room. Figure 5.7 shows a plan of a traditional lounge space with what I would describe as furniture that is 'unengaged' architecturally with the building form. Windows are an obvious omission for the purposes of this illustration.

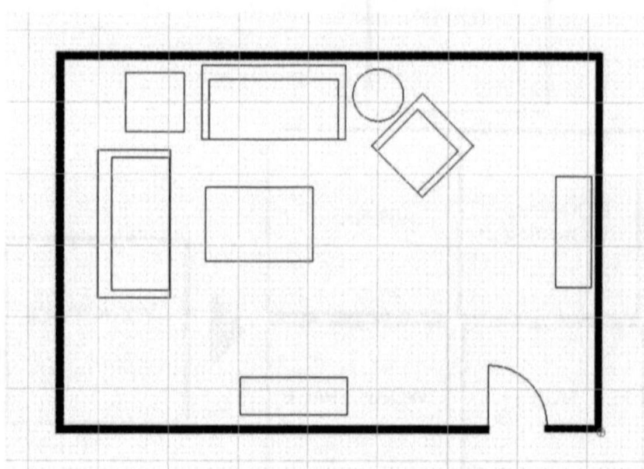

Fig. 5.7 Conventional plan of a lounge area showing loose furniture that has no place of ownership

5.2.1 *Planning Using Both Cells C and D from Fig. 4.24*

Rectangular Footprint—157 m^2
Total floor area—218 m^2

Using the cell D in Fig. 4.18 as the main storage/activity provision with the rectilinear extensions works perfectly for all spaces. Cell C from Fig. 4.18 is used for the utility room and the circulation/storage area. This area will have open storage under the stairs. The cells do not retain the cell as an area entity but merely serve to be adapted within the grid. Note the lower hinged child stable doors that can be used to seal off the hub area to a toddler. All splayed angles are windows (Fig. 5.8).

Problems:

- Utility storage under worktop awkward access.
- Accepting the lounge needs a three-way seating arrangement this space does not have enough peripheral storage.
- Hall cramped.
- Stair well has too much circulation.
- The external configuration does not make for a terraced or semi-detached solution.

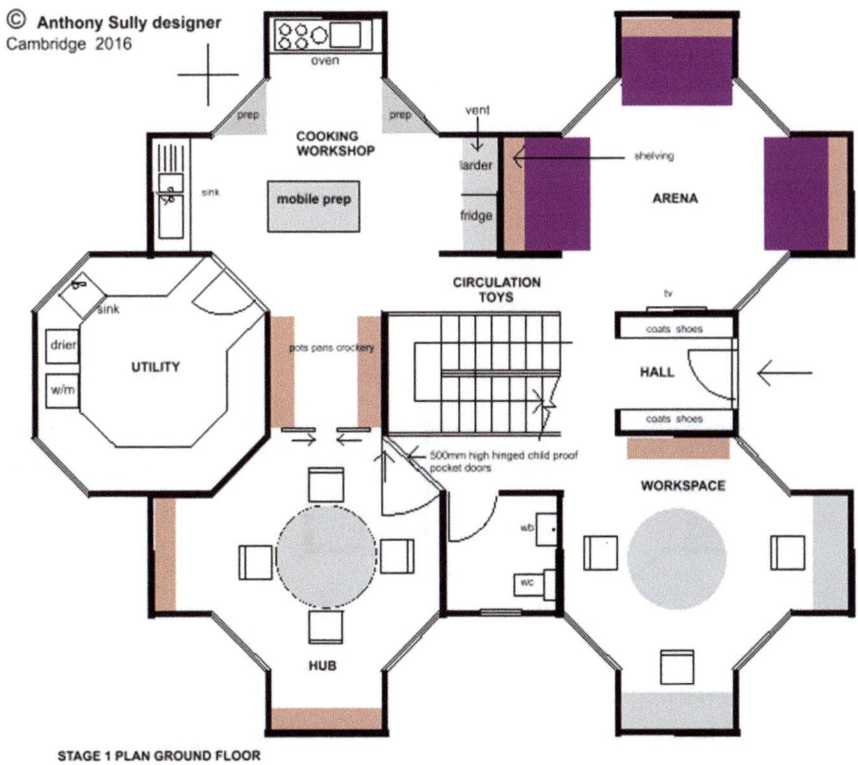

Fig. 5.8 Plan of ground floor

5.2.2 Planning Rationale

- All external angled planes are windows giving good all round views from every side of the building but not floor to ceiling.
- Play area sealed off by folding/childproof doors and first floor have sliding pocket doors.
- Utility area with sink caters for washing, drying, ironing, household maintenance.
- Study for two people.
- Dogleg staircase NOT straight staircase.
- Stairwell for storage and pushchair.
- Flexibility of change of use of cells according to occupier's needs.
- Furniture can be built-in or supplied by occupant.
- Despite compartmentalisation spatial vistas can be experienced.

The octagon is the generating form in enclosing the interior spaces but half of the octagon is experienced externally around the perimeter three times and a quarter of an octagon is experienced four times (Fig. 5.9).

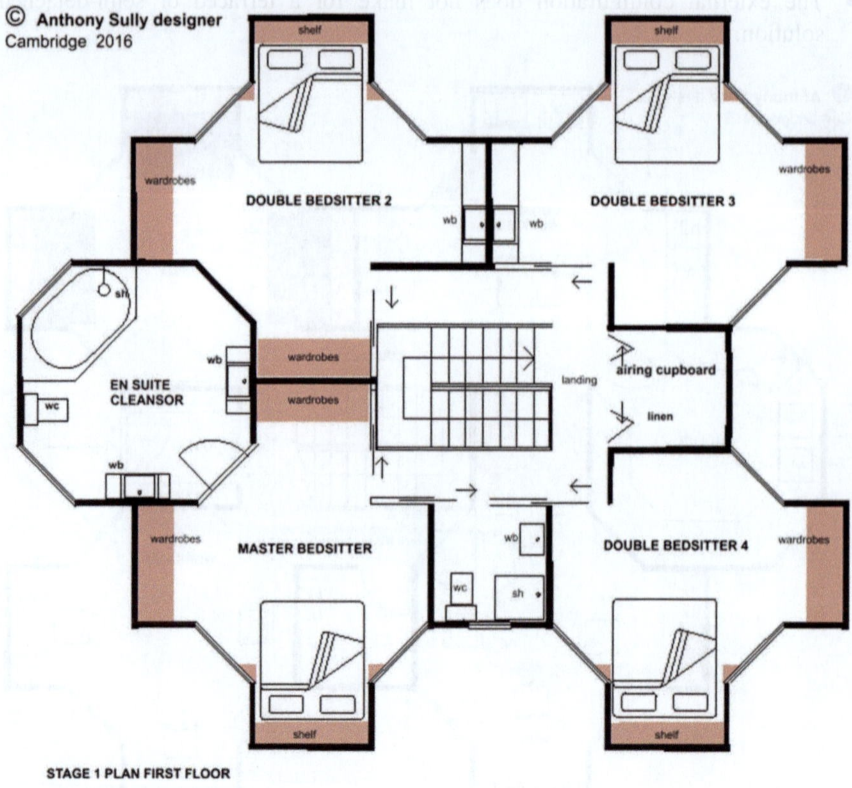

Fig. 5.9 Plan of first floor with four double bedrooms, including one en suite

Two bedsitters have their own basin. Ample storage provided for linen. Beds slide out for access.

Problems:

- Four double beds are too many and each rather more spacious than need be.
- Bed recesses too deep.
- Too much circulation around stairs.

The first floor has a maximum capacity of one master bedsitter with en suite cleansor and two double bedsitters with shower and basin in the room, and one double bedsitter. In addition, there is one separate wc, basin and shower room. A good-sized airing/linen cupboard compared with the almost non-existent space for such in many housing schemes that are built. Each bedroom has space for additional furniture such as chairs, a desk and dressing table about which the occupant will decide.

5.3 Stage 2—Design Proposal 2—Preferred Solution

The intention is to reduce the area of the previous scheme and produce a more compact solution that can effectively compete with current house designs on the market. This version can be built as

- Detached, conjoined or terraced—all double width frontage.
- With a choice of three-roof designs.
- The third-roof design: the footprint of the ground floor remains unchanged when accommodating two bedsitters, three bedsitters, four bedsitters or five bedsitters by simply building over the existing ground floor structure.
- Two house sizes proposed: total internal floor areas—146 m^2 (Large) and 126 m^2 (Small).
- Includes single den and bike/bin store.

5.3.1 Planning

I have taken the cellular compartmentalisation of Stage 1 as a beginning and merged the cells to produce the right-hand plan in Fig. 5.10. I have used a combination of four of the larger cells per floor and merging them on the ground floor whilst omitting the smaller cell. The only remnants of the octagon left are the chamfered walls in four corners of the house. Maintaining the stair as the central core to a double fronted concept the front and rear entrance fell into place. The resulting plan form reduces the chamfered wall presence to only four compared

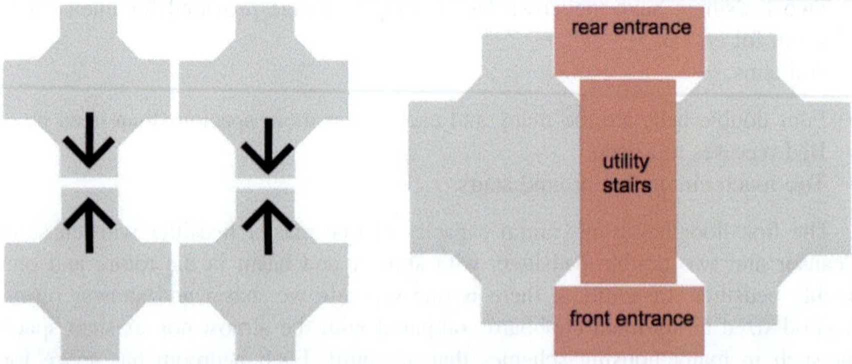

Fig. 5.10 Merging of cells from Version 1. Left-hand diagram shows direction of merger. Right-hand diagram shows how the rectangular core blocks are added

with 16 before but without losing any of its effect. The remaining forms perform the 'enclosing' function and use the rectilinear grid to rationalise size and position.

Basic Area Relationships for a Three-Bedsitter House
Fig. 5.11 shows the central spine almost becomes automatic consisting of hall, stairs, utility and toilet. This creates a duality in the planning: a left and right and a front and back. The cooking workshop links strongly with the dining area (hub) which can change to a protected safe play space for toddlers. This is a recognised need when the parent is busy in the cooking workshop or utility areas. The utility operates well adjacent to the cooking workshop as well as being close the rear exit. The other side of the house has the arena space connected to a workspace with flexibility of adjustment depending upon particular family needs. The toilet is much better placed at the rear and is more discreet rather than close to

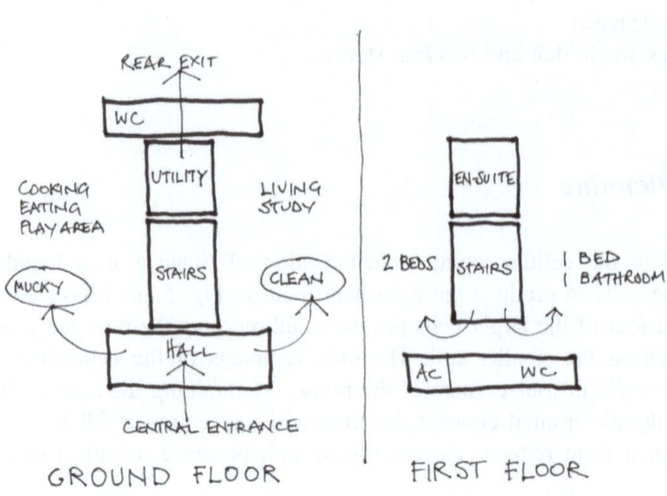

Fig. 5.11 Planning concept for ground floor: taken from Fig. 5.6. Division into Mucky side and Clean side

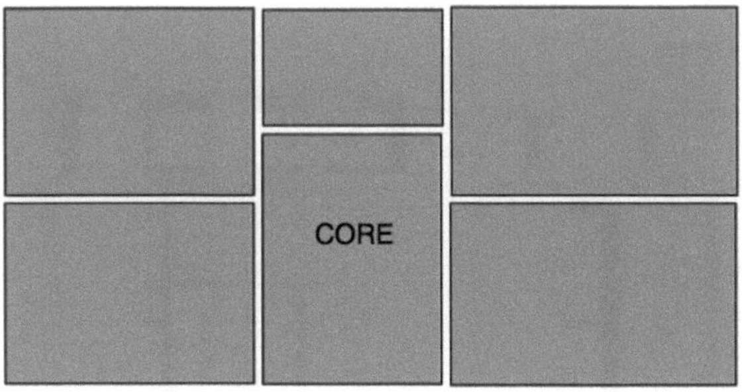

Fig. 5.12 Block diagram of spaces around the central core

the front hall as is most common practice in house design. The division between 'clean' (arena and workspace) and 'mucky' (cooking workshop and hub) suggests itself right away. Direct links from the cooking workshop to the den and utility are also considered desirable.

5.3.2 Translated into a Building Form—Larger Area of 146 m²

This is the three Bed version with adjoining den. It also has 11 windows in total as opposed to 24 in Stage 1. All windows are now in the contained rectangle and the chamfered walls are solid. See from Fig. 5.33 onwards the 3D form with a variety of roof designs.

- Footprint area without den—102 m^2

The generative portion of the plan in Fig. 5.13 shows the principle of the rectilinear 'contain' function shape combining with the angular 'welcome' face.

Fig. 5.13 Generative key concept plan

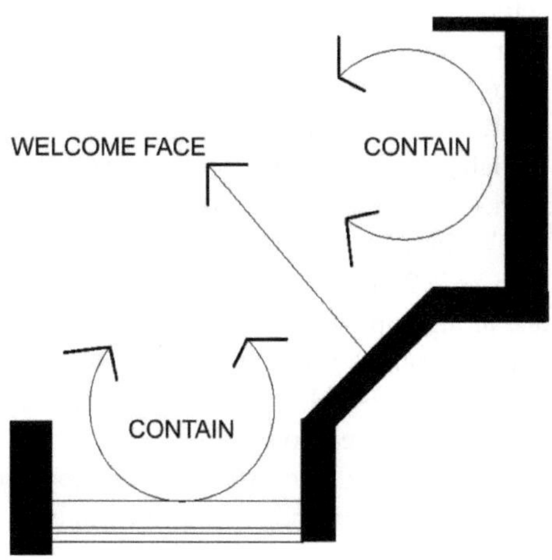

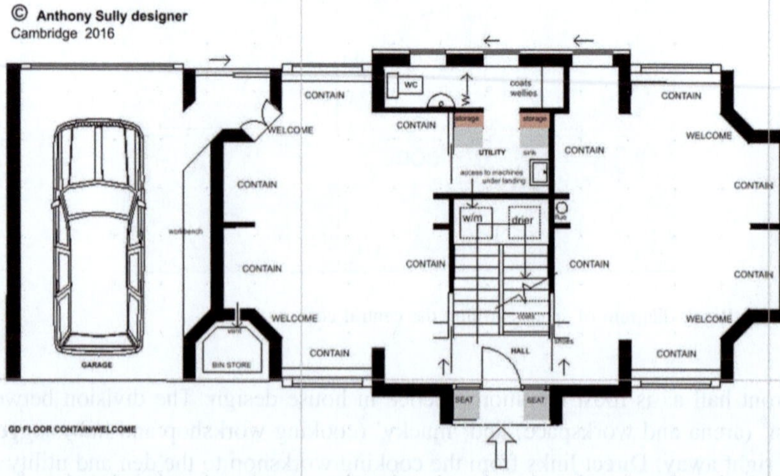

Fig. 5.14 Showing the contain and welcome parts allocated on the ground floor plan. See the ground floor plan fully developed in Fig. 5.15

Labelled as the C and the W. The C contains storage and work surfaces at a depth of 600 or 300 mm and the W destroys the dominance of the square corner that exists in every house. The W is therefore more welcoming, is the 'plain' wall of the house, and can be used for decoration, mirror, message board or any other fixture. This concept is not part of any existing house planning doctrine used by architects or developers that I know of. Figure 5.14 shows how the proposed plan is formulated with the C and W separate from the core blocks.

The plans in Figs. 5.15 and 5.16 show a more realistic external wall thickness of 250 mm. The relationship of kitchen, dining and lounge has been varied judging by the case studies in this book with open plan being favoured containing these three activities. As I have stated previously, this popular solution has to cope with overlapping functions that can clash having a destabilising effect on the people involved. It has become a plan of economic necessity blinded by the false impression of wealth that large spaces tend to create. I am proposing an option to partially open plan the cooking workshop with the hub and the arena with the workspace with low-height moveable partitions.

The first floor has the master bedsitter/en suite combo on one side of the house, with beds 2 and 3 and cleansor on the other. The central core has the separate WC and airing cupboard off the landing. There is a corridor/storage space by bedsitter 3, which provides a possible access to two more rooms that can be built over the garage (den) (see Figs. 5.56 and 5.57). Bedsitter 2 is the only room that does not have a chamfered corner.

The front and back windows are opposite each other twice on the ground floor. The cooking workshop and arena windows can hinge open upwards (see Fig. 5.17)

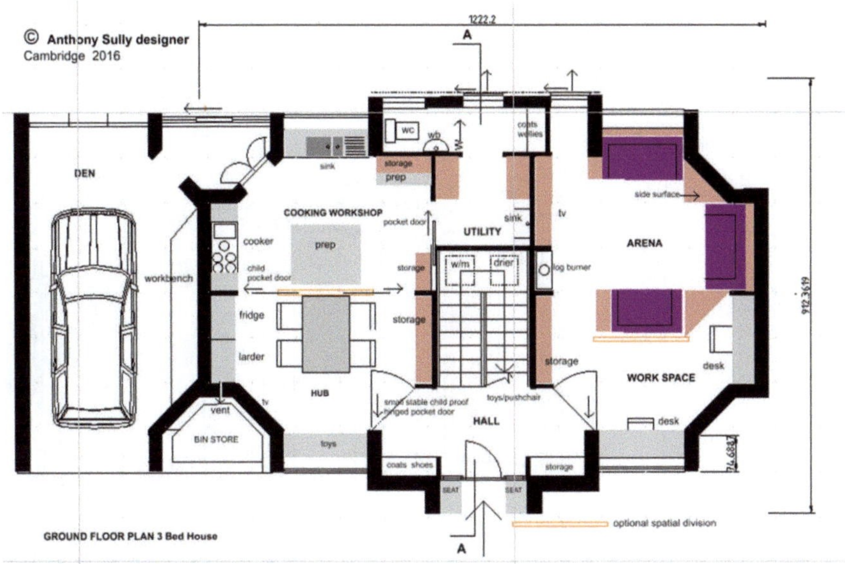

Fig. 5.15 Ground floor plan

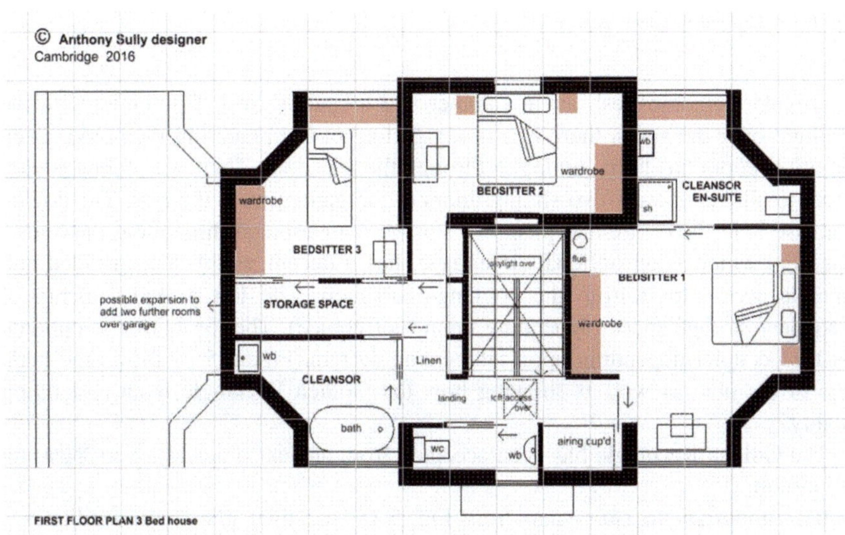

Fig. 5.16 First floor plan

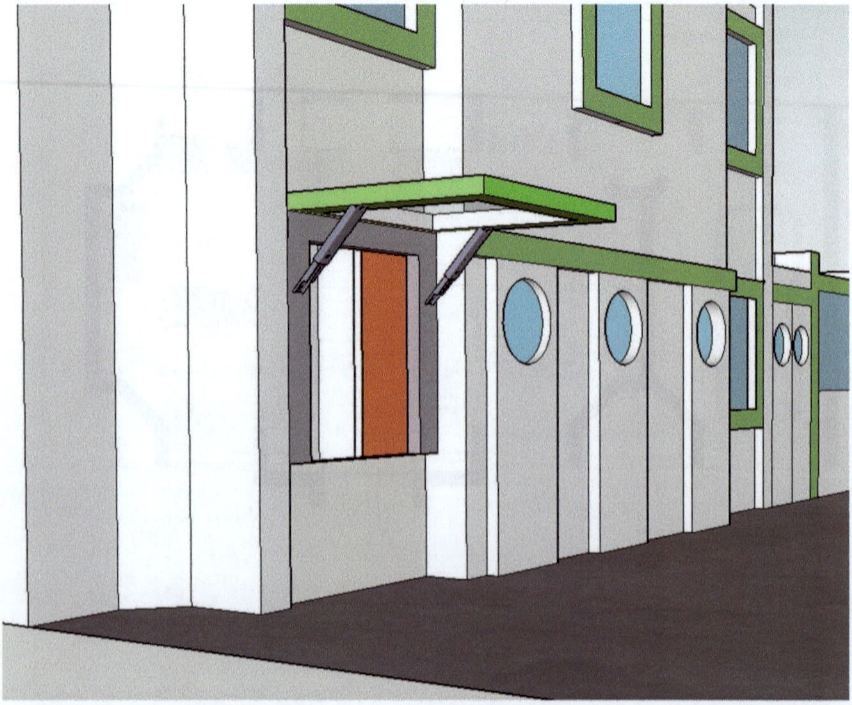

Fig. 5.17 Hinged window upwards

to provide welcome fresh air in summer and a canopy over. Combined with the rooflight over the stairs, there is a strong feeling of light entering the house. Even the utility room has light coming in through the rear door. There are sliding pocket doors to all rooms with two sliding doors to the garden from the rear. The dogleg staircase is in the centre and plenty of storage is distributed throughout the house. The den, which is an optional appendage, has a decent workbench installed and there is access to it from the cooking workshop. The loft provides plenty of secondary storage space (depending upon roof choice). The log burner is optional for the occupier depending upon the heating system. Figure 5.19 shows the main axis of the plan as well as showing how the Golden Rectangle is an underlying quality.

The Golden Rectangle has been accepted from ancient times as a very pleasing proportion.

It is important to stress that from Fig. 5.15, the final plan form facilitates a visually recognisable means of adjoining a den or adjacent property as shown in Fig. 5.20 and as can be seen in the conjoined houses of Fig. 5.36.

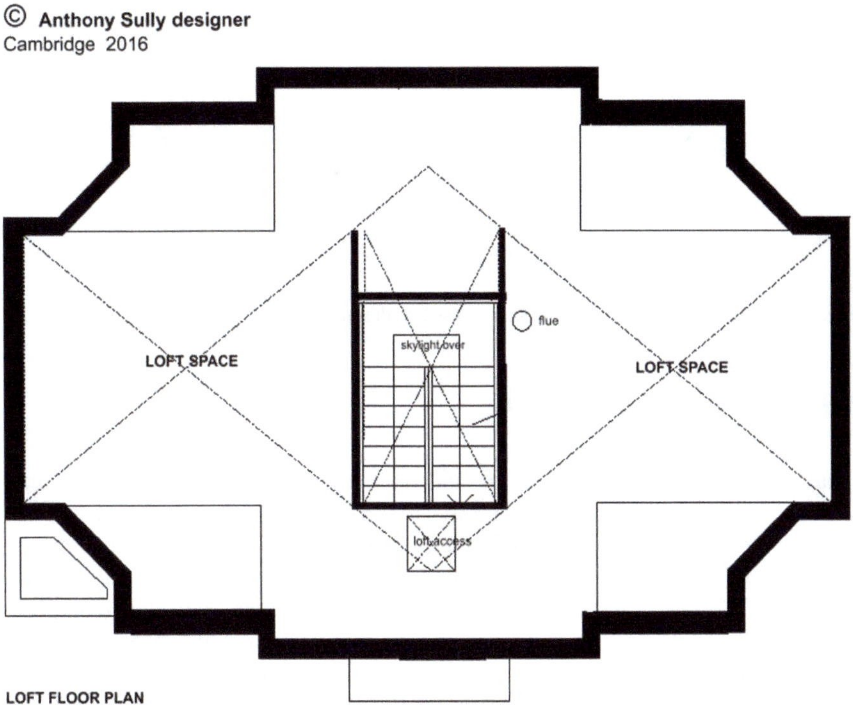

Fig. 5.18 Plan of loft that fits roof design in Fig. 5.35

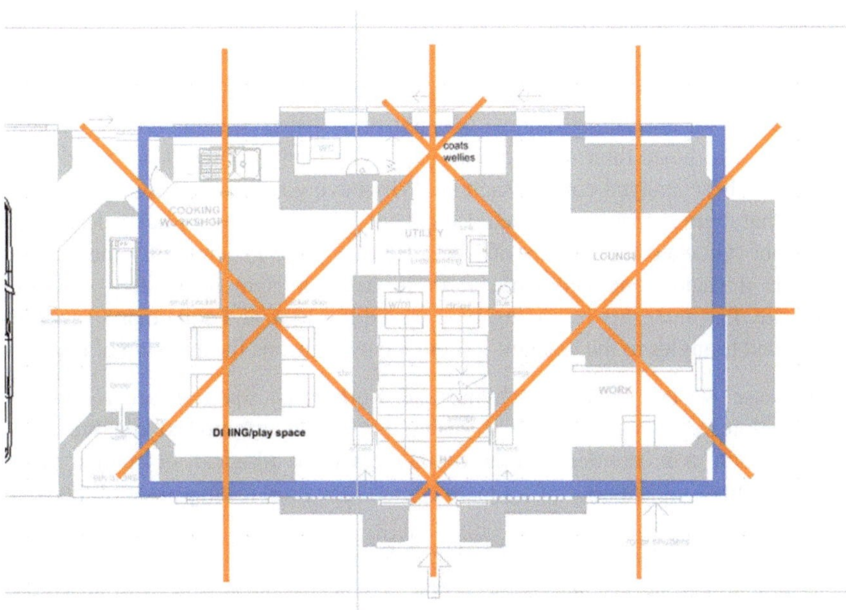

Fig. 5.19 Main planning axis plus Golden Rectangle in blue

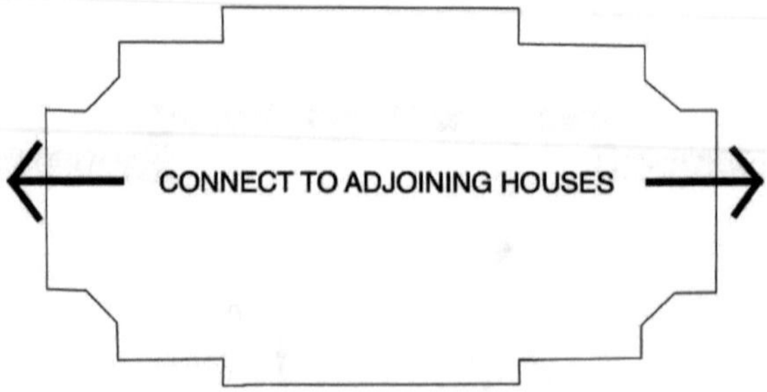

Fig. 5.20 Sides designed to adjoin other houses or den neatly

5.3.3 How Activities Work

I have placed the washing machine and drier under the landing of the staircase hence providing more space in the utility area for laundry and ironing. The hub also doubles up as a play space for children and has a ventilated larder. It also has the fridge/freezer, which, in my experience, serves this hub equally as the cooking workshop. The cooking workshop, hub, arena and workspace have clear boundaries and separate storage. There is an option to install a low or full height partition to partially separate each of these joint spaces should the occupiers so desire it. The London's Housing Design Guide of 2010 recommends for a three-bedroom four-person house to have a joint area of the living, kitchen and dining areas to be 27 m, and my small scheme here has 47 m^2. The staircase is still a central circulation core lit from the skylight above and provides storage for coats, shoes and pushchair on the ground floor.

The arena space dispenses with the ubiquitous central coffee table by providing adjacent surfaces (arena sides) for drinking vessels. The discipline of being specific about a place for everything reduces the anxiety of guests of being unsure where to sit or where to place their drink as can happen in many houses. I have had some reaction to these splayed corners such as 'Why have you knocked the corners off?' hinting that there is space to be used if I had not. The first thing to make clear is that I have not 'knocked the corners off' because they were generated by the octagon which is a stand-alone geometric form in its own right as readers will now appreciate. I counter this by saying 'you would not look at a square and say' they have added triangles to the octagon'. There are two reasons for this kind of reaction:

(1) Because we are so used to living and working in square-based rooms, the directional pull of the walls is so strong that any interruption to this directional pull, such as a chamfered corner, will seem to be illogical, or counter to the 'normality' and stability of a square corner—the final resting place of the walls meeting.
(2) The chamfered wall is seen as being reductive to the interior space compared with a square corner, and some people will feel a sense of loss of space that

should be rightfully theirs. The strength of this context is that the chamfer connects with a rectangular recess (container) rather than other walls continuing the octagon.

In order to defend my plan shape against the square I have reproduced, as an example in Fig. 5.21, only the cooking workshop and hub space as a rectangular plan that has exactly the same floor area as my original splayed angle plan.

I have not added corners to the chamfered plan to make the rectangle as that would increase the floor area. The rectangular area must be the same as the original and it is quite clear that this rectangular box has no character whatsoever. I have indicated by the red question marks those unsatisfactory places, which have a

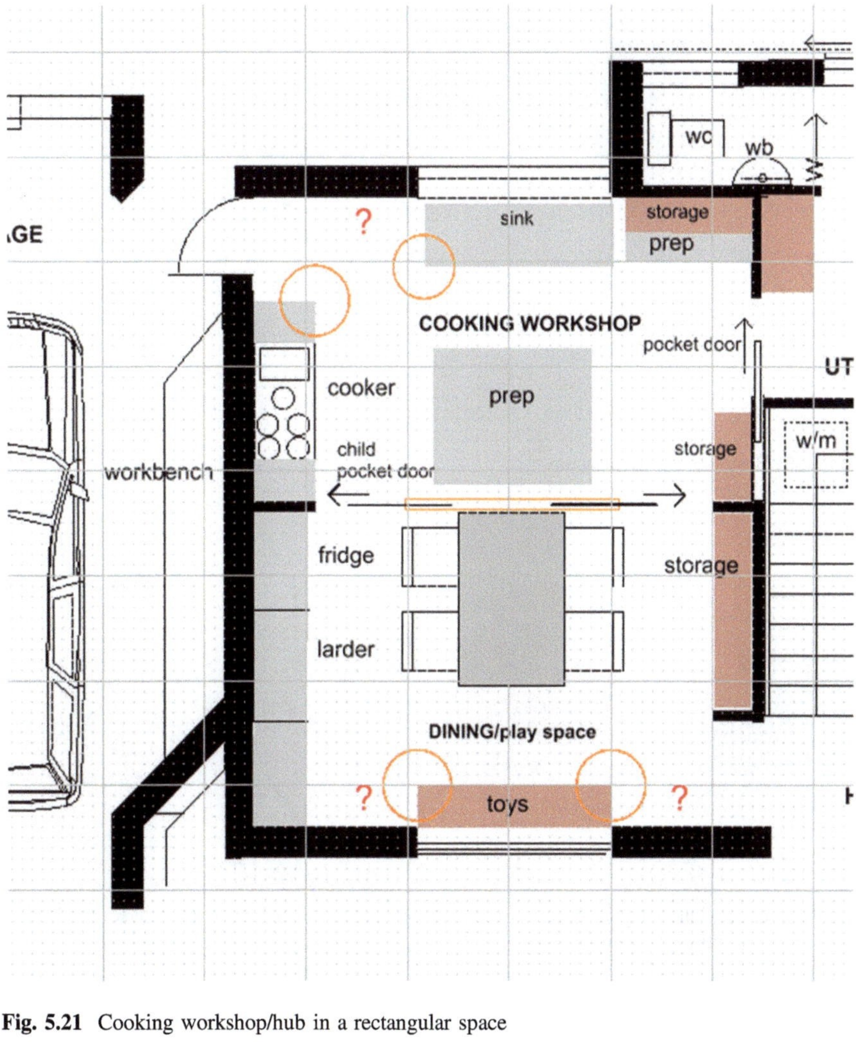

Fig. 5.21 Cooking workshop/hub in a rectangular space

leftover nothingness. The red circles reveal unwanted intrusive corners and the door to the den is stuck in the corner. This plan resembles Fig. 5.7 whereby the fixtures (apart from the existing ones against the core staircase) have no home.

The basic structure shown in Fig. 5.22 goes through both floors and is made up of two component types that gives the building a strong integrity, and this is what is lacking in many house designs today (Fig. 5.23).

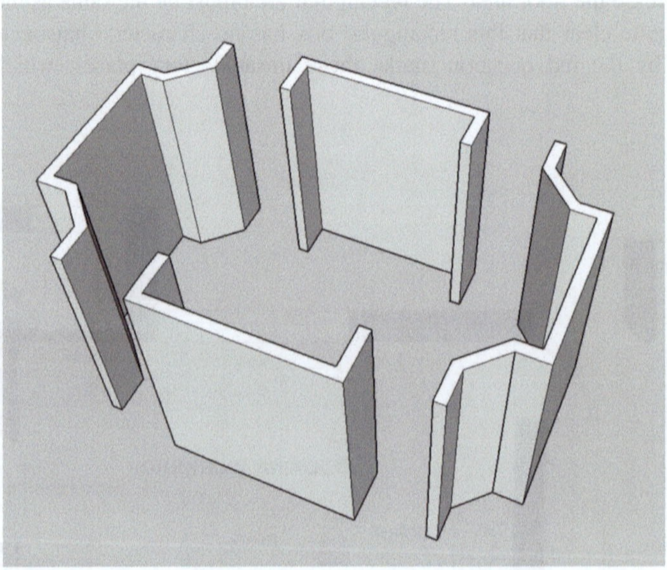

Fig. 5.22 Basic solid wall structure

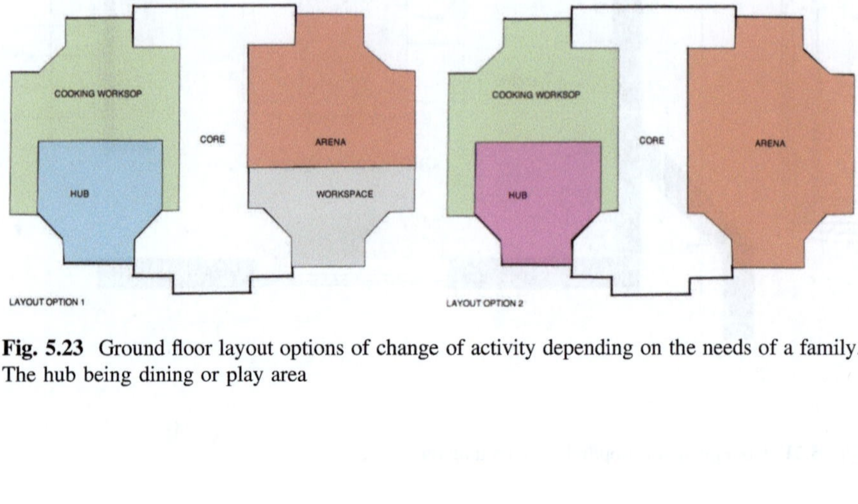

Fig. 5.23 Ground floor layout options of change of activity depending on the needs of a family. The hub being dining or play area

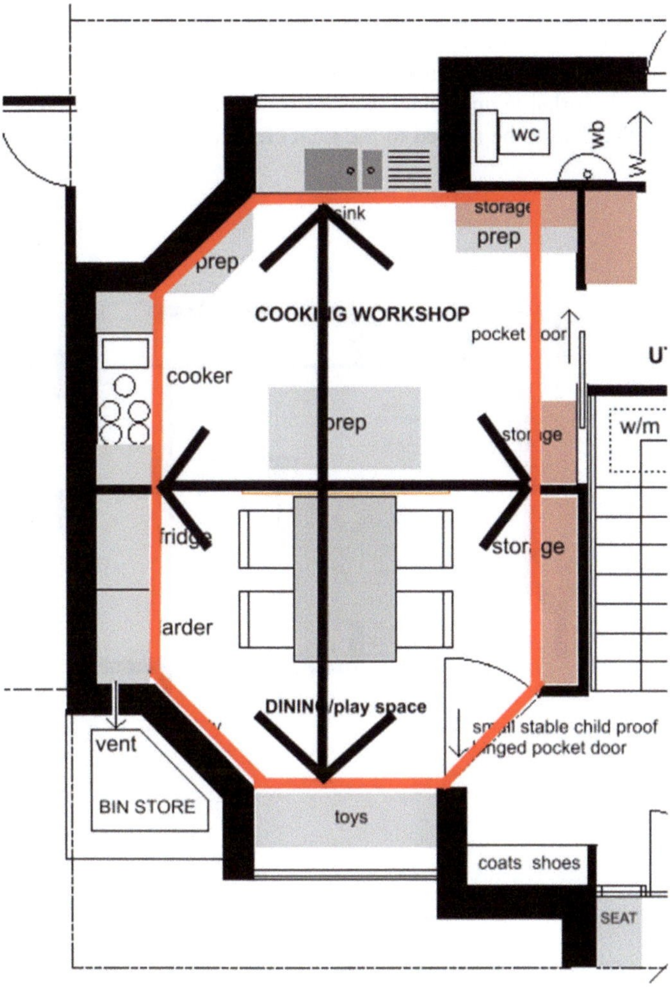

Fig. 5.24 Visual boundary

Figure 5.24 is half of the floor plan which shows the circulation space in red as being the real visual boundary of the interior as opposed to the 'contained' fitted furniture/product items behind the boundary.

5.3.4 Reduced Size of Stage 2—Three-Bedsitter House

- Two singles and one double bedsitter
- Footprint of house without den 77 m^2
- Total internal floor area 126 m^2 (excluding den)

Here is a smaller scheme (Figs. 5.25 and 5.26) designed to compete with existing house designs in the marketplace. I have reduced the width dimension of the internal splayed walls by half which results in a slight reduction in all areas but the plan is still workable and relationships do not change. With a reduced size of landing, the WC had to be moved to the joint cleansor and the airing cupboard has moved.

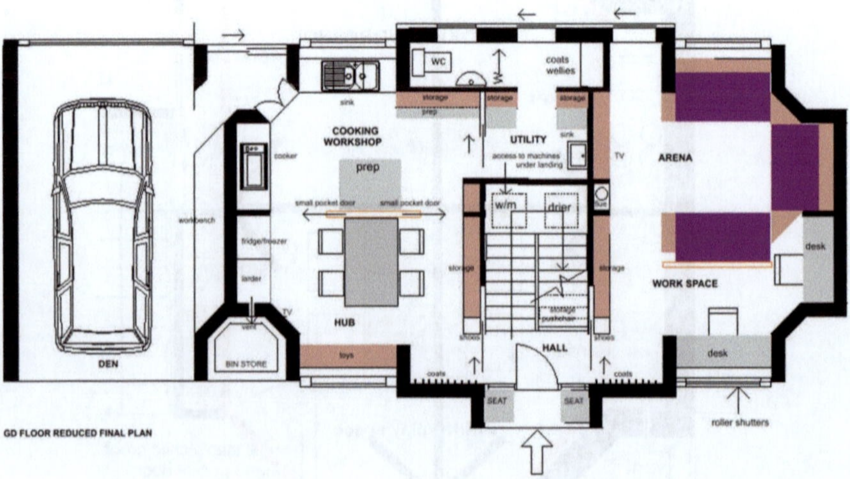

Fig. 5.25 Ground floor plan

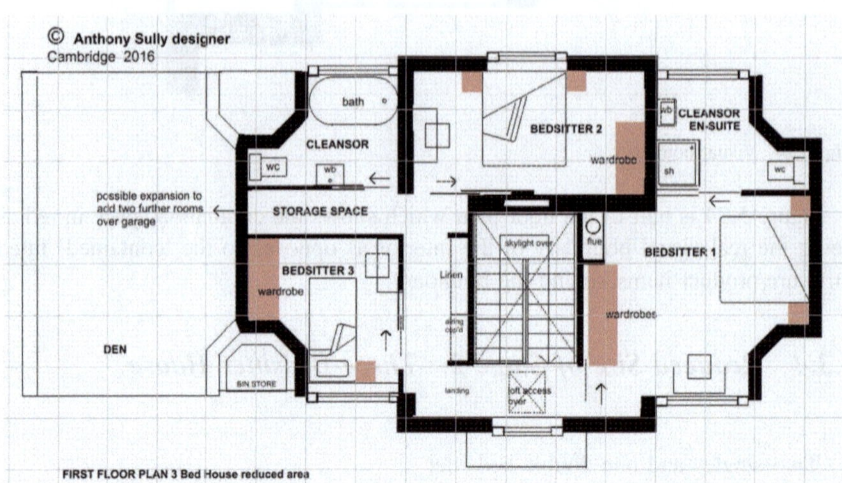

Fig. 5.26 First floor plan

For this first floor plan, I have switched bedsitter 3 with the cleansor so as to group the water/waste services on both floors towards the back of the house. The latest RIBA consultative report of May 2017 'Homewise—Space Standards for Homes'[1] disappointingly still sticks to 93 m^2 as being the minimum internal floor area for a three-bedroomed five-person house. The fact that this design version is what I consider to be the minimum means that there is a serious misunderstanding by all consultative bodies on what the needs are for any activity in the home. This scheme is the preferred solution because it more realistically competes in area with existing house types that are built and also offers more flexibility in estate planning.

Important Interior Vista Considerations that Help Determine the Space

It is important that upon entry one does not feel cramped which is most common with many entrance halls. Even with the sliding doors closed, the stairwell is a welcome escape. These views 'join' the two halves of the house (Fig. 5.27).

This shows how each activity has its own working space as well as the views through the windows (Fig. 5.28).

It is important that the full depth of the house is experienced (Fig. 5.29).

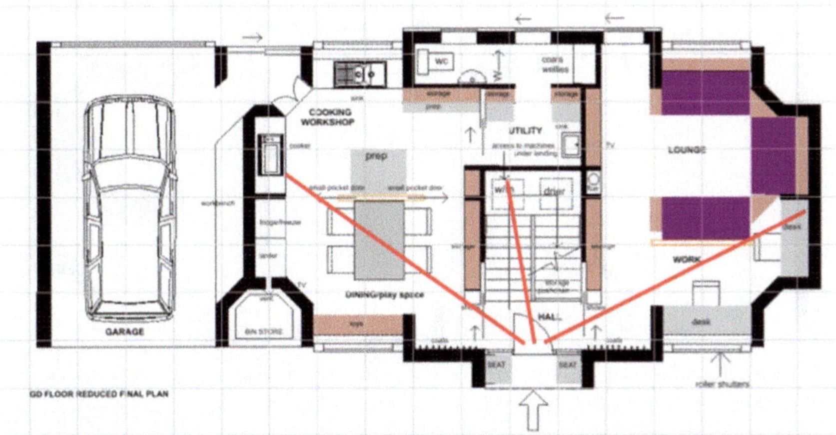

Fig. 5.27 Entry vistas

[1]RIBA consultative report of May 2017 'Homewise—Space Standards for Homes'. New homes still not big enough.

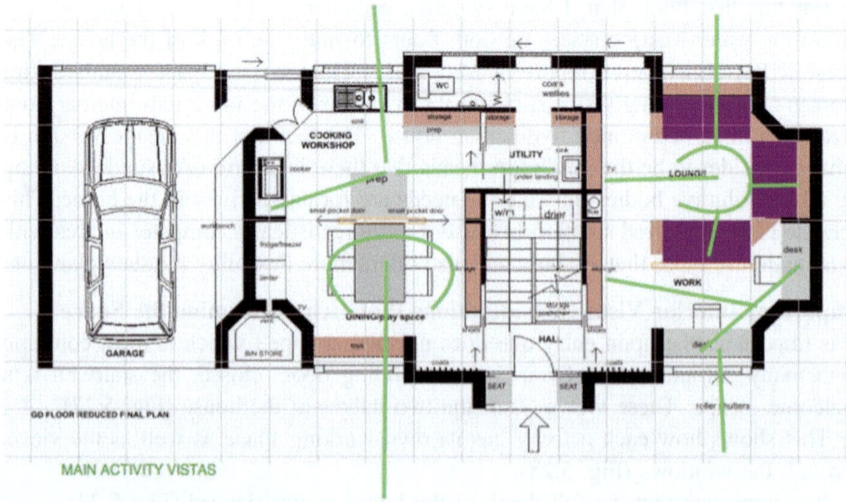

Fig. 5.28 Main activity vistas

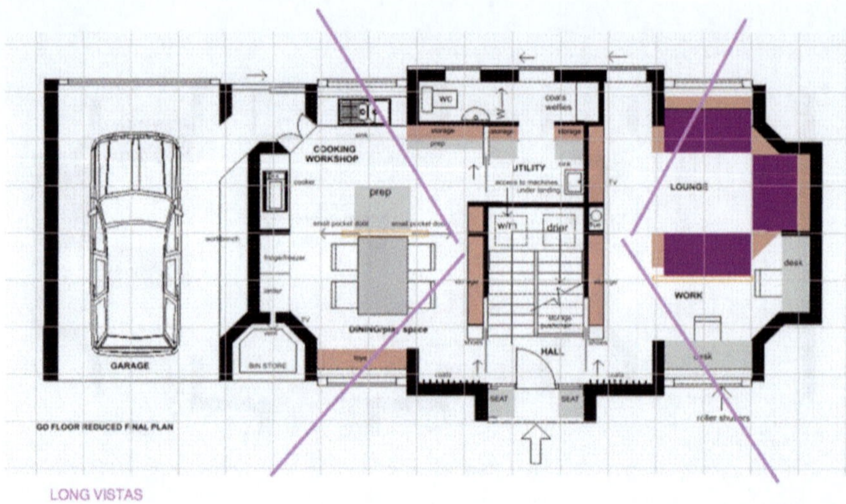

Fig. 5.29 Long vistas

5.3.5 3D Interior, Exterior, Section and Elevations

The exterior form in Fig. 5.33 has a pitched roof and gabled ends, which includes a skylight over the staircase. The walls are less undulating than Stage 1 with only 10 windows throughout and hence create a simpler front elevation. The house has a wider frontage than most of the case studies, which are usually narrow front and long depth. The roof form is not directly dependent upon the interior of the house but rather on its exterior and interior wall structures whilst acknowledging some form of rooflight over the central stairwell. This design I consider to be the more traditional solution. Hence, I have proposed three more concept variations of the roof design in addition to the standard pitched roof to demonstrate what could be done. See Sects. 5.3.6, 5.3.7 and 5.4 (Figs. 5.30 and 5.31).

Please note that the major ground floor spaces either side of the core stretch from front to back of the house giving a good sense of space and light. It is so important to break out of 'rooms' and provide vistas. Light also floods into the hallway from the front door plane as well as the skylight above the stairs. The value of the skylight is to provide natural ventilation as well as daylight to the stairwell. All windows to have roller steel shutters for security, as is more common in France and Germany. I cannot understand why this is not standard practice in the UK considering the number of house burglaries we have (Figs. 5.32 and 5.33).

The Windows

In Sect. 1.7.4, I mention that the windows in my proposed design are strongly expressed and as can be seen in Figs. 5.34 and 5.35 and they have wide pronounced coloured frames. These are also echoed in the interiors. Because of the generative diagram of Fig. 5.13, they have a fixed position in plan, which is shown on the elevation as far as the width positioning is concerned. The height position is

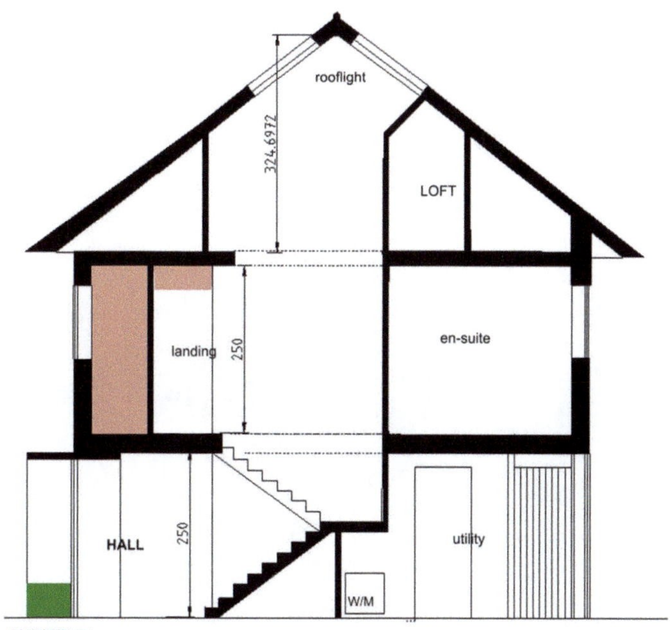

Fig. 5.30 Section AA through building from Fig. 5.15

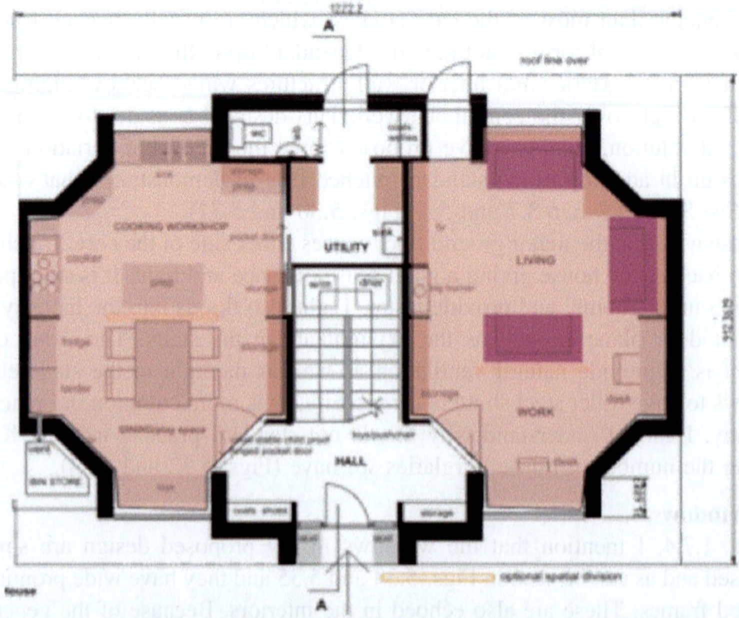

Fig. 5.31 Showing single large spaces shaded

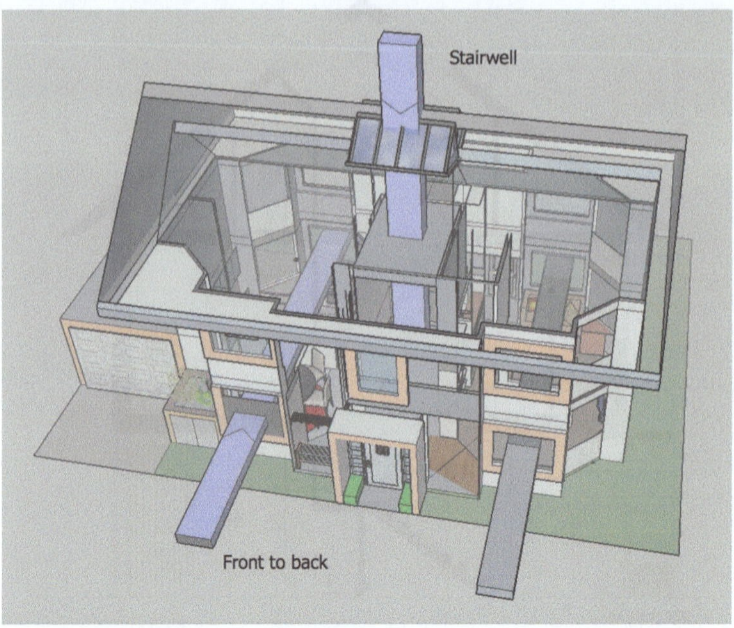

Fig. 5.32 Drawing showing main daylight through lines

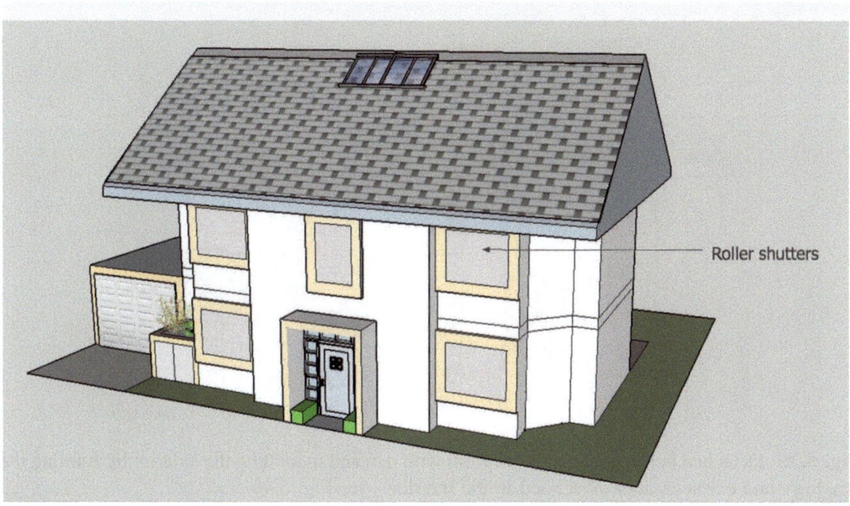

Fig. 5.33 3D view showing window roller shutters and pitched roof

Fig. 5.34 Front elevation

determined by the 900-mm internal cill height and the abutment of the frame against the ceiling. The frames are large enough to incorporate the security metal roller shutters.

As I have explained in Sect. 1.2, Figs. 5.36 and 5.37 show a conjoined pair of houses.

The front porch provides the option of seating and useful storage for parcel deliveries, wellies, etc. It also satisfies one of the recommendations by the RIBA (Sect. 3.1.2) of having a welcoming atmosphere as well as providing shelter to postmen or anyone

Fig. 5.35 Detached house gable ends proposal with den and modelling the exterior by framing the windows and entrance door as echoed in the interior—see Fig. 5.40

Fig. 5.36 Front view of conjoined houses

requiring a signature when raining. Figure 5.38 shows a traditional design of a sawn green oak porch made today by Hartwood Oak Buildings in Berkshire.

Figures 5.39 and 5.40 are indicative of how furniture can be fitted into the house but variations are possible depending upon the user. The following drawings show how many variations are possible for the roof design (the hat) whilst still retaining the 3D form (body) of the building. The case studies show three projects that have innovative roof designs that integrate with the structural form more than just being a 'lid'. See Sects. 2.3, 2.3.6, 2.3.7 and 2.5.7.

Fig. 5.37 Rear view of conjoined pair

Fig. 5.38 Brick and handmade sawn green oak porch frame under a clay tiled roof by Hartwood Oak Buildings

Fig. 5.39 3D view of interior of ground floor

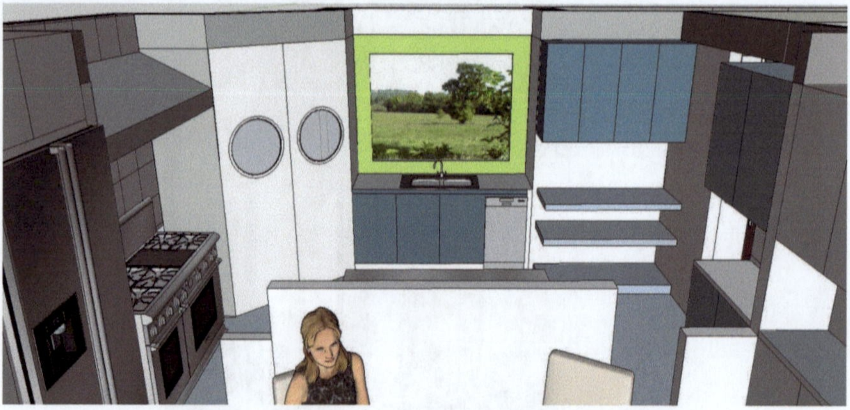

Fig. 5.40 View of hub area with cooking workshop in rear. Note childproof pocket doors (see plan Fig. 5.25) and internal window framed as the exterior is

5.3.6 Alternative Roof Design A

This proposal makes the roof shape acknowledge the chamfered corners of the plan and reduces the heaviness of the roof, but in doing so loft space is sacrificed and a more complex roof structure is required (Figs. 5.41 and 5.42).

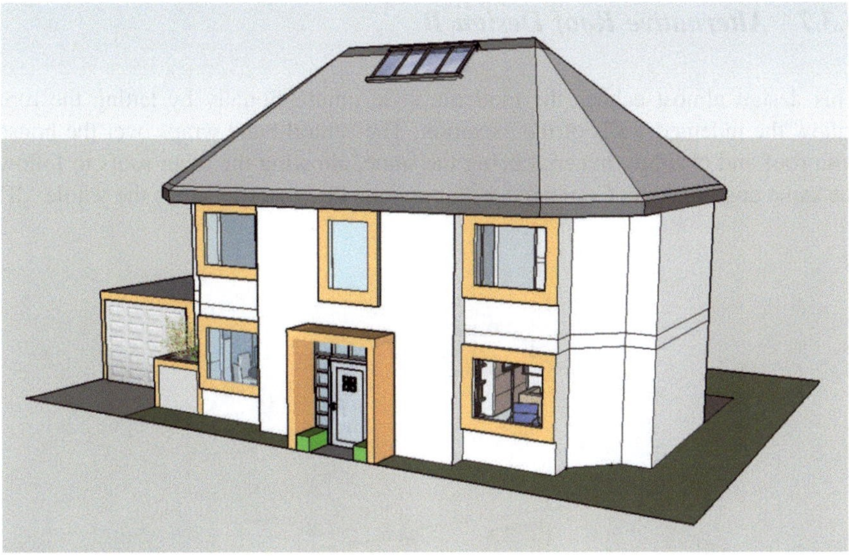

Fig. 5.41 Hipped roof design

Fig. 5.42 Front elevation of hipped roof design

One huge difference between the more common rectangular house plans and my design proposals is how adjustment of size is described or achieved. With rectangular houses, size is described by the number of bedrooms: i.e. two bed, three bed, four bed in that the ground-floor-serving spaces increase proportionately, whereas my design proposal in Sect. 5.4 presents a two- or five-bedroomed house using the same size ground-floor-serving areas (Figs. 5.43 and 5.44).

5.3.7 Alternative Roof Design B

This design almost echoes the moderne style unintentionally by letting the roof follow the indented walls of the elevation. The central band wraps over the house with roof and cladding material being the same, allowing the other roofs to follow the same angled slope. Compared with previous designs, this alters the whole '3D

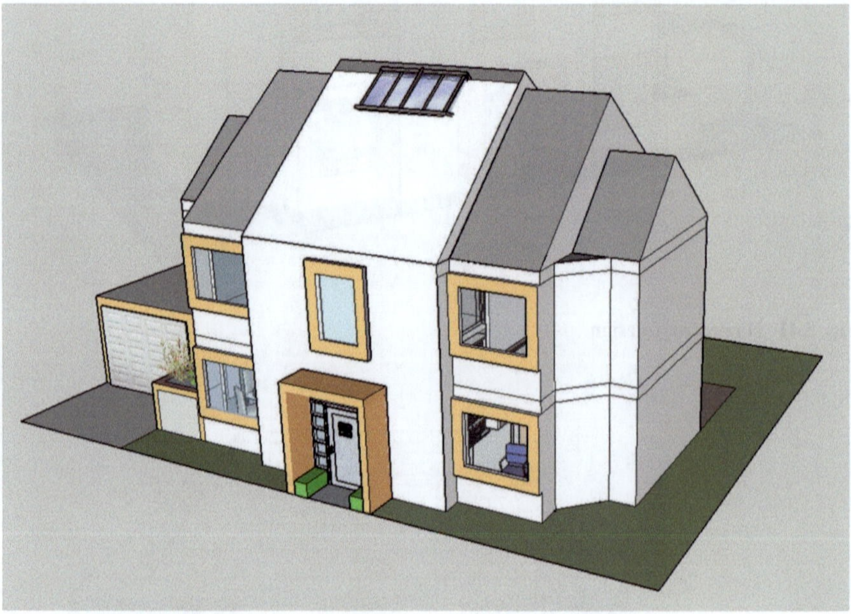

Fig. 5.43 Stepped roof design

Fig. 5.44 Elevation of stepped roof design

effect' and is not just a variation of a 'lidded roof'. The loft has height reductions but still follows the plan. The chamfered corner creates a small flat roof. This design is more honestly expressive of the 3D form of the building.

Confession

The den in all the previous designs looks like an unrelated appendage that has been stuck on. Section 3.2.4 has already shown how horrendously out of scale a double den is to most houses. This is not uncommon with many house designs in the last century, as the car appeared from the wings to announce its arrival as an uninvited guest to an architectural evolutionary party. Domestic architecture has ever since found the problem of hosting the car's arrival to the house, its parking and ultimately storage, to be a headache as can be seen in Figs. 1.25 and 2.1. Figure 5.45 shows one solution for a fairly new luxury house by admitting defeat in combining the den with the house, and instead a separate building is built in the style of the main house.

As a result of this conundrum, the following roof design has enabled me to link the single den with the house more strongly as can be seen in Fig. 5.56. But this design has not been conceived to just suit the two- or three-bedsitter house as you will see, but it also plays an important role in further house expansion.

Fig. 5.45 Large house with separate den. Wikimedia Commons, geography.org.uk. Author Brian Shaw

5.4 Alternative Roof Design C—On 126 m^2 Floor Plan Including Terrace

I have chosen this design to demonstrate how from the same ground floor plan a two-, three-, four- and five-bedroomed dwelling can be built. IT IS NOT NECESSARY TO BUILD ANY GROUND FLOOR EXTENSION FOR THESE EXPANSIONS! Up to now, families have had to move to a larger dwelling or build a ground floor extension to accommodate their expanding family. This proposal rotates the axis of the pitched roof 90° which results in a gable end at the front and two flat roof areas either side resulting in a much smaller loft storage area. The flat roofs will drain to internal rainwater pipes. The entrance porch echoes the gable end roof. The chamfered corners are more boldly expressed by allowing them to rise above the flat roofline evoking ancient castle castellations. For plans, please see Figs. 5.25 and 5.26. Here are four first floor plan diagrams illustrating each bedroom combination (Figs. 5.46 and 5.47).

Fig. 5.46 Two-bedsitter house showing roof design C with a pitched roof porch and a more integrated den than previous designs

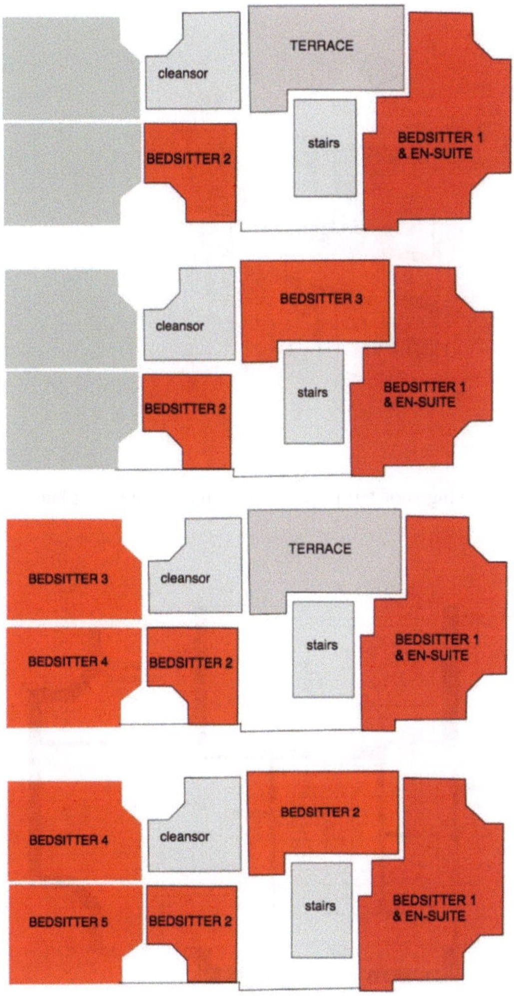

Two bedsitters (Fig. 5.49)
Includes terrace.

Three bedsitters (Fig 5.26)
Terrace is used for 3rd bedsitter—simple add on.

Four bedsitters
Bedsitters 3 and 4 built over den leaving the terrace open for use.

Five bedsitters (Fig 5.55)
Maximum occupation. Terrace is built over for bedsitter 2.

Fig. 5.47 Four plan diagrams of each first floor bedsitter combination

5.4.1 Two-Bedsitter Version—For Front View See Fig. 5.46

- One double and one single bedsitter (Fig. 5.48)

Ground floor plan as in Fig. 5.25.

The roof terrace is where the third bedsitter will be otherwise the plan stays the same. The terrace can be served by the mini kitchen adjacent to bedsitter 2 (Fig. 5.49).

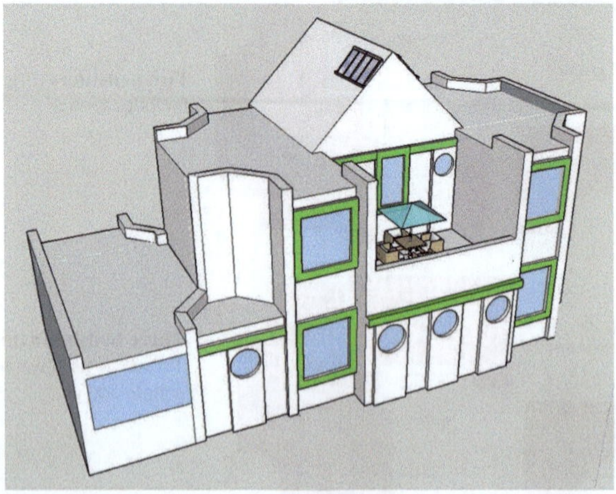

Fig. 5.48 Rear view of two-bed house showing roof terrace and den handed from the plans

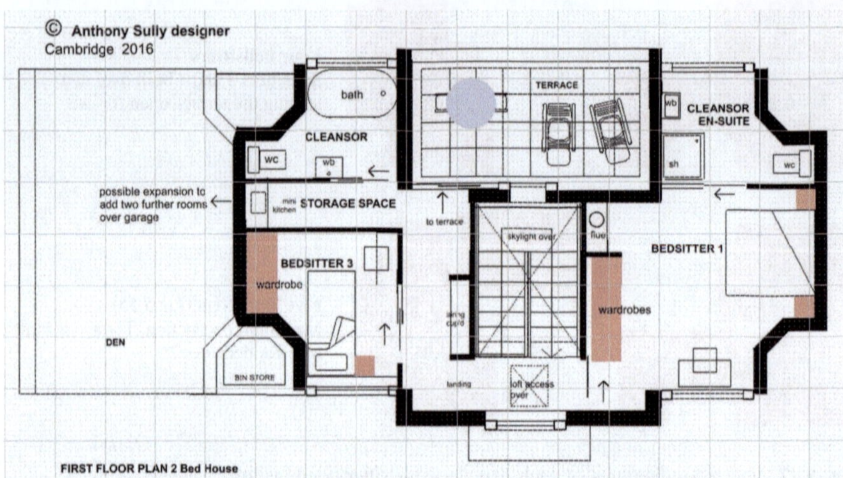

Fig. 5.49 First floor plan showing roof terrace

5.4.2 Three-Bedsitter Version

This shows that the den can work on either side of the house in which case the plan layout would be handed. Please see Figs. 5.25 and 5.26 for plans of both floors. To change from a two bedsitter to a three-bedsitter dwelling, the first floor terrace is built over to the edge of the building continuing the existing roofline (Figs. 5.50 and 5.51). A four-bedsitter dwelling can be achieved by building the two bedsitters over the den and allowing the terrace to be used (see Fig. 5.55).

Fig. 5.50 Front view of three-bedroom house

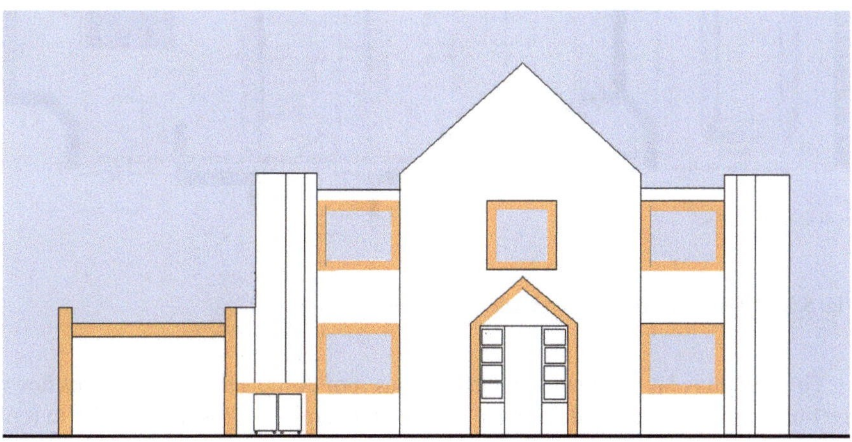

Fig. 5.51 Front elevation

How does my design proposal compare with an existing estate house? Figure 5.52 shows an overlay of an existing estate house elevation, and Fig. 5.53 shows the plan overlay, and one can see that there are similarities. Mine has to be slightly larger to satisfy my call for more space (Fig. 5.54).

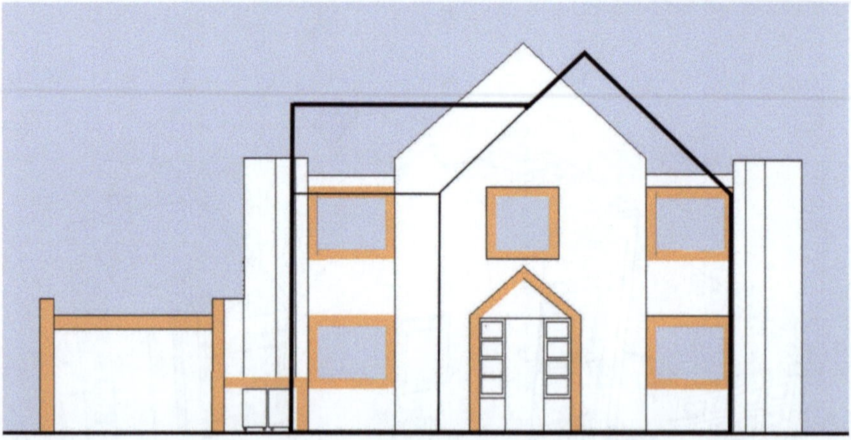

Fig. 5.52 Overlay of standard four-bed house from an existing estate house

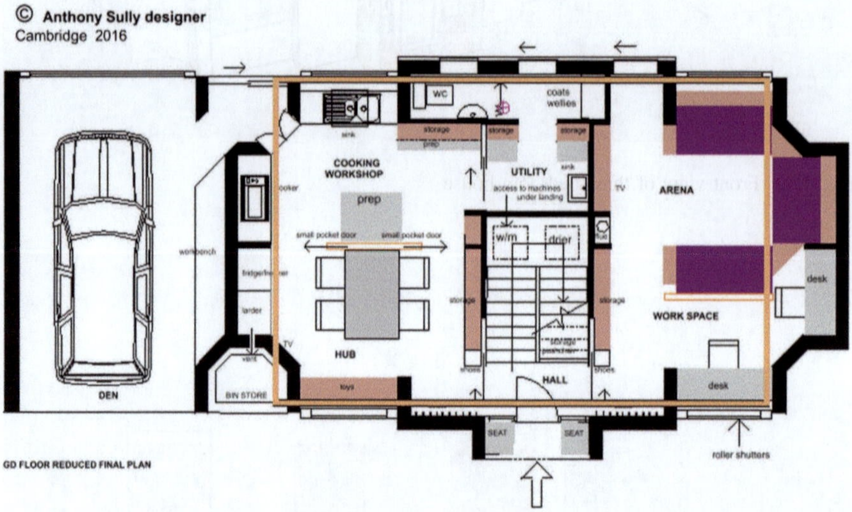

Fig. 5.53 Overlay in orange of an estate house ground floor plan

This design is far stronger than the previous roof designs because it breaks down the building into visually identifiable component parts. Yet it is also similar to roof design B in that it is not a variation of a 'lid' but presents a totally different 3D solution, even though in each case the plan remains the same. There is a rhythmic indentation that binds the den with the main building rather than looking as though it has been tacked on. The validity of these proposals will depend upon the method of construction and ultimately cost (Fig. 5.55).

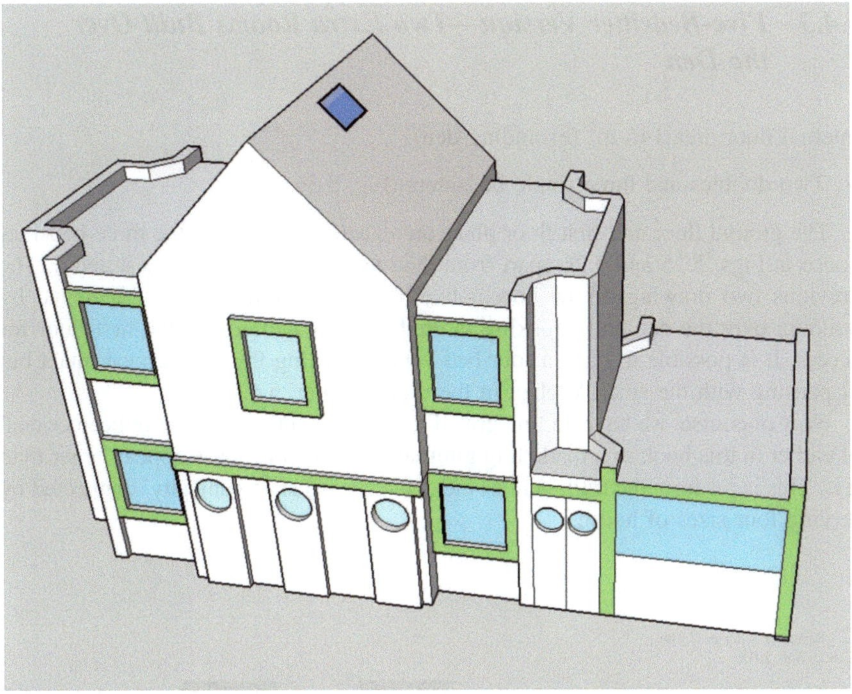

Fig. 5.54 Rear view showing two sliding doors to the garden. Double doors to den/cooking workshop

Fig. 5.55 Showing two three-bedroom houses conjoined

5.4.3 *Five-Bedsitter Version—Two Extra Rooms Built Over the Den*

Internal floor area 146 m^2 (excluding den)

- Two doubles and three single bedsitters (Fig. 5.56)

The ground floor and first floor plans are exactly the same as the three-bedsitter house in Figs. 5.25 and 5.26 apart from the two additional bedsitters 4 and 5. The previous two drawings show how a five-bedsitter dwelling can be achieved by building over the den and knocking through the storage area on the first floor for access. It is possible to have a four-bed version leaving the roof terrace intact but dispensing with the small kitchen in the corridor (Fig. 5.57).

Now of course, we have an 'integrated' den a description that I have been critical of earlier in this book as a marketing gimmick to make the house appear larger than it is. This is far from the truth here as the den has existed 'organically' connected by serving four sizes of house.

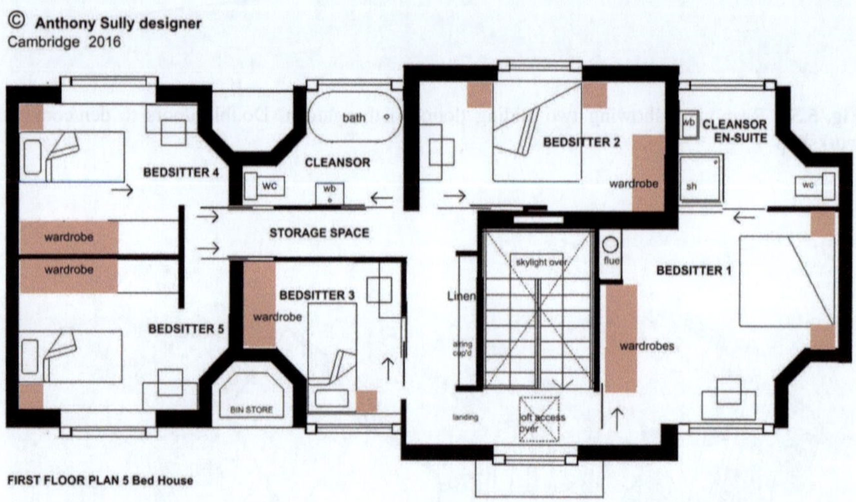

Fig. 5.56 First floor plan

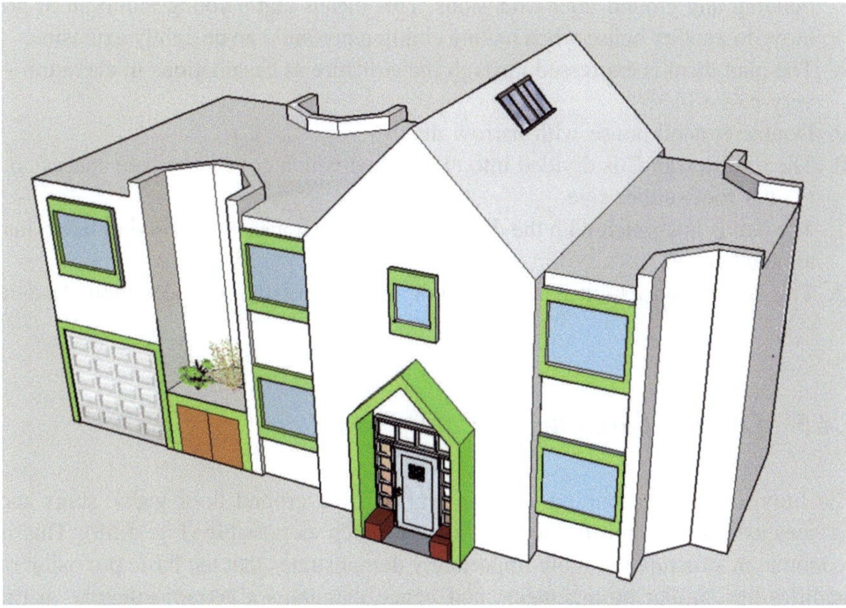

Fig. 5.57 Front view of five-bed house

5.4.4 Summary of Design Intent

Interior

1. To minimise the square corner and replace by chamfered walls that are a welcoming factor and are based upon the use of the octagon in early planning ideas. Using Generative diagram Fig. 5.13.
2. To eliminate the fashionable floor to ceiling glazing as being a sales gimmick rather than offering anything practical. Window cills are great!
3. Dogleg stairs instead of straight stair.
4. Hinged doors replaced by sliding pocket doors. Low height ones for making the play area (dining) safe for toddlers.
5. To provide more space and storage throughout the house contrary to standard regulations.
6. To re-name spaces in line with present-day living.

Exterior

7. Ground-floor-serving spaces should be the same size whether there are two or five people living in a house with a single garage (den). On that basis, the two-bed scheme has the same footprint and can expand to three, four or five beds without

building any ground floor extensions. This means that a young family need not move to another house when having children nor build an unsightly extension.

8. The plan form is expressed through the structure as castellations in elevation—very castle like!
9. Double fronted house with narrow depth.
10. The roof design C is divided into (1) pitched which covers the core spaces and (2) flat roofs either side.
11. The den is integrated with the design of the main house as opposed to being just an appendage.
12. This design can be built as detached, conjoined (semi) or terraced without the den.

5.4.5 *Layered Planning on Each Floor*

The fully integrated planning solution works if the ground floor walls, stairs and services extrude through to the first floor as much as possible (Fig. 5.58). This is economic in structure but more importantly demonstrates that the basic provision of facilities has similar arrangements and hence maintains a certain integrity in the design. Historically, the majority of houses have followed this principal except when we come to the Barratt House in Sect. 2.3.2 whereby the ground floor has nothing in common with the first floor apart from a common exterior structure. Indeed it is even more common to find a disparity between the different floor plans of many other building types which weakens its 3D design integrity.

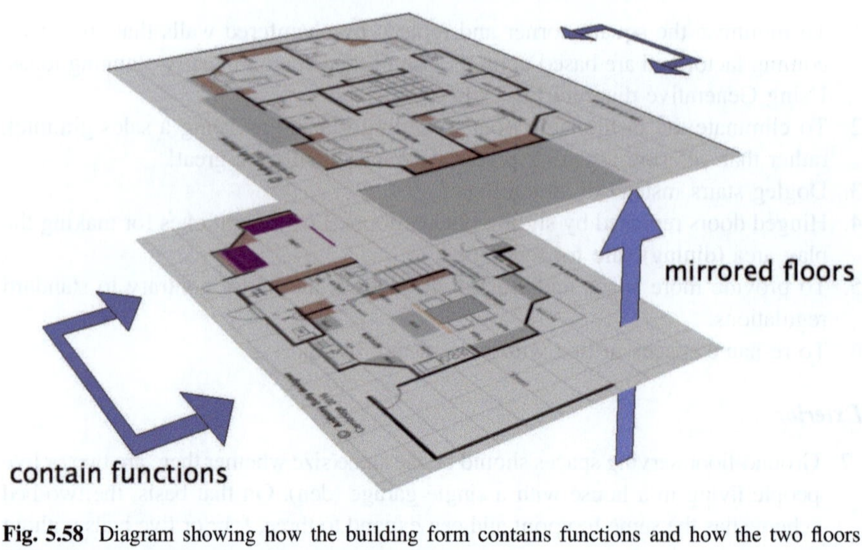

Fig. 5.58 Diagram showing how the building form contains functions and how the two floors fairly mirror each other as an integrated 3D solution

5.5 Estate Planning

5.5.1 Access

When dealing with access to the property and its rear garden/terrace, it seems that the options are as I have detailed in Fig. 5.59. Whilst I have concentrated on the design of the house in this book, the success of any design is dependent on how well it can be planned to form an estate development. Of course, there are many complex schemes that have been built that explore the landscaping and mix of dwellings to create interwoven structures that are innovative and visually exciting. This book is not dealing with such issues but I feel the need to show how my proposal can work at the simplest level.

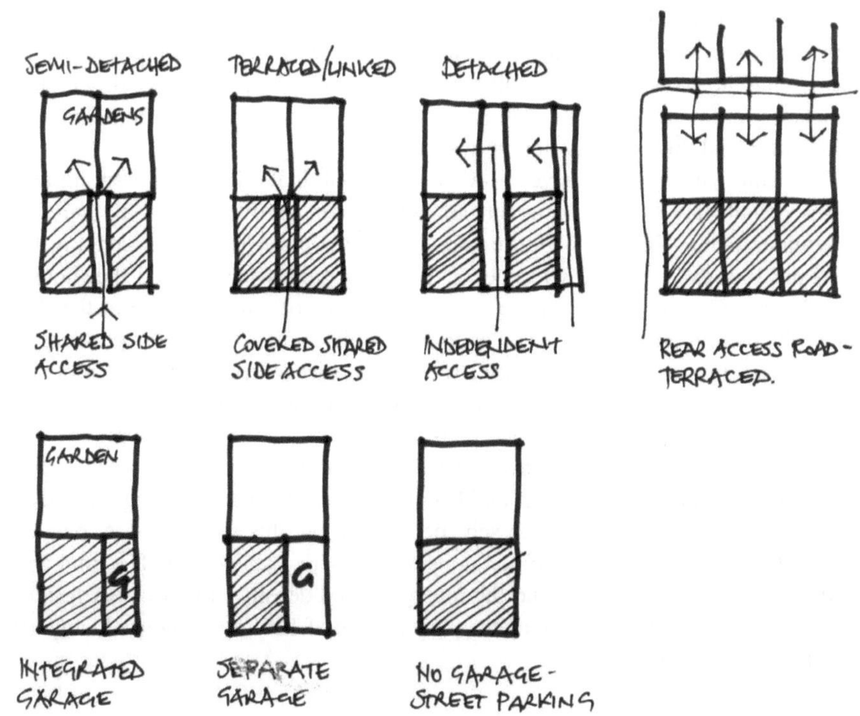

Fig. 5.59 Sketch plans showing access to rear gardens/terraces

5.5.2 *Plot Size*

This comprises the amount of land around the house, known as the curtilage, upon which the house and single den sits allowing pedestrian access front and back, side access to rear garden, and parking for one car on the driveway. The second car is in the den. Regulatory dimensions have been used such as 6 m frontage from house to pavement, and 11 m depth to the garden. This gives a plot size of 374 m^2. Regulations should be made to prevent owners turning their front garden into a parking area.

Please remember that this concept proposal has grown from the inside out using a single figure to plan the interior spaces together with the generative geometric shape in Fig. 5.13. As I have said before in the book, the house should be designed as a product and nobody is doing this. We have a system in this country which has been publicly condemned at a conference I went to recently, (especially compared with the system they have in the Netherlands) whereby the machinery of government together with the cartel of the large building developers, and Local Councils, have to take collective responsibility for the mess the whole housing industry is in. Much work is being done with regard to planning communities and providing the right sort of infrastructure for them to survive, and the building technology field is awash with new products aimed at satisfying the goals of sustainability and energy conservation. But we desperately need the government to realise that the family home is being replaced by tense pig-pen architecture, and this is surely not the way to go.

The main concern that I have shown in this book is the lack of storage in new houses, which has led the occupants to use the den as overflow. It is therefore incredulous to me that many houses are built without a den, hence denying the occupants any extra storage whatsoever. Planning guides recommend that houses should have off-road parking space for two cars, one of which can be in the den. The problem with the integrated den as shown in Fig. 5.59 is that it puts pressure on the designers to design a three-storey structure in order to keep within the plot size (see Case Study 2.3.2). The separate den would seem the preferred choice. I am not in favour of grouping neighbourhood dens together as one way of reducing the plot size, because they can be too far away from the house for working in, and also they can end up not being looked after, and becoming a visual eyesore. The plan shown in Fig. 5.61 shows my Stage 2 scheme (based on the single house plan in Fig. 5.60) laid out with 12 conjoined houses each with a single den. The total area for the above plan is 5170 m^2 and the recommended density of houses per hectare is 30. This means that with my reduced plan version of 126 m^2 per house I can only get 24 houses in one hectare.

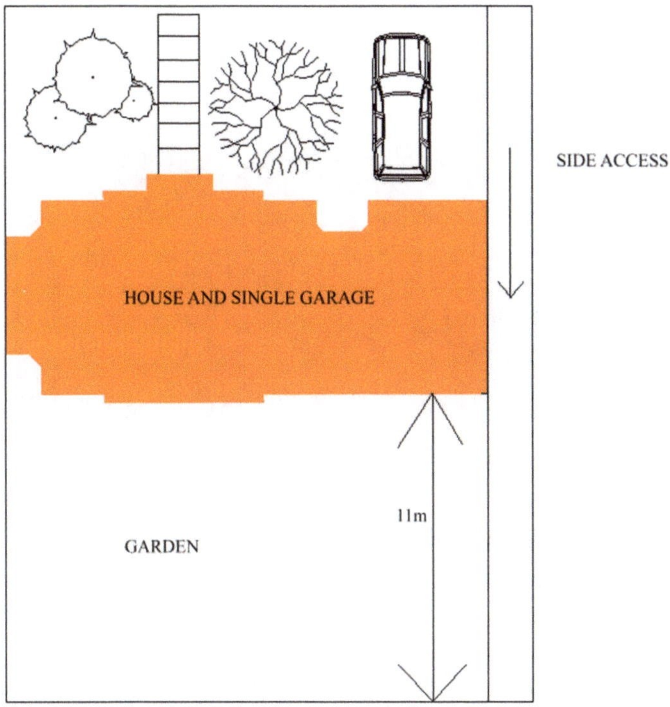

PLOT SIZE 374sqm

Fig. 5.60 Plot size derived from the Stage 2 house design

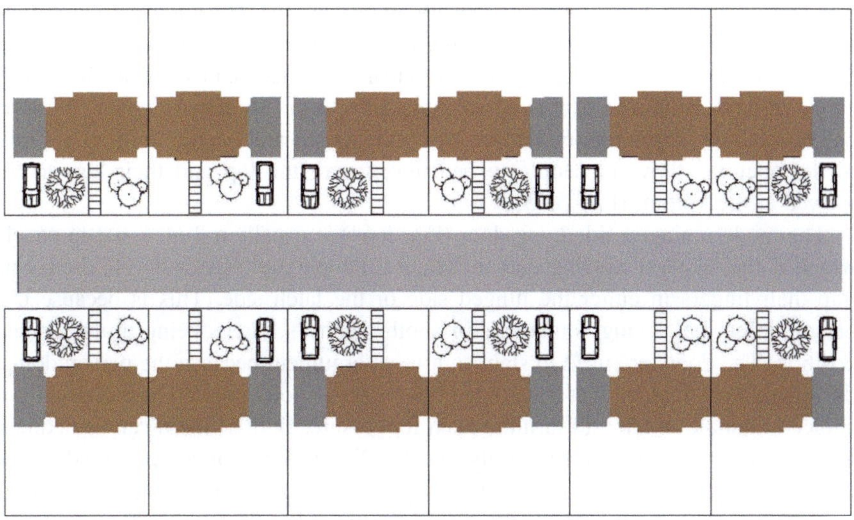

Fig. 5.61 Example layout showing 12 houses with a road of Stage 2 design as conjoined with adjoining dens and side access to rear garden

Fig. 5.62 View of sample estate with conjoined houses

5.6 The Case for the Sliding Pocket Door Instead of the Hinged Door

Nobody knows when the hinged door was invented but to begin with it was only used as a front entrance to a building and not in the interior. When an opening was left in the primitive structures made with branches and clay or the hide of an animal the instinct would have been to make a removable panel of same material. Eventually, the thought of tying it to one side with twine would be made, thus creating the hinged action. It could of course be tied above the opening but the side is easier to manage. From then onwards for all of these centuries in every country in the world the hinged side hung swinging door has been used. Of course there are cultural variations such as sliding doors in Japan, or veiled/beaded fabric hung from above in Arab countries. The most amusing ones to me are the American Western cowboy saloon double doors that are two-way hinged and occupy only two-thirds of the opening allowing people to be viewed over the top and their legs to be viewed at the bottom (Fig. 5.63).

The modern hinged side hung door (Fig. 5.64) is usually a 2 m × 0.8 m panel, which is the heaviest moving part in a house. People and especially children, can trap their fingers in either the hinged side or the latch side. This is because the person doing the closing is unaware of another party's fingers being in the danger position. The doors are used to slam shut in an argument because the person doing the shutting has learned that a slammed door makes a loud noise. There is usually a choice of handle or knob to turn the latch for closure. Both of these can sometimes get caught in clothing, especially the handle. When the door is open (and doors usually open into the room) it is often left open and if so, it is an unwanted intrusion into the room and can disturb the room's setting. Its existence when closed is

Fig. 5.63 American Western Saloon entrance double swing doors. *Drawn* Author

primarily to continue the plane of the wall on the room side. On the other side it is set back from the wall surface by the door frame.

When opening the door to gain access to the room it is assumed that nothing will impede the swing action on the room side, but how often have little toddlers been the unfortunate recipients of such action, or even adults carrying a tray being crashed into? Yet these doors continue to exist because nobody has really applied any design thinking to them. The sliding pocket door as Fig. 5.65 shows, solves all of the above problems. When closed it is recessed back from the wall surface on both sides of the wall to allow for its sliding action back into the thickness of the wall. When it is open it disappears and no longer becomes an interruption in a space. It disappears! There are no projecting handles to get in the way. The design of the mechanisms and recessed handles has improved greatly over the years. The wall construction must be thick enough to receive the thickness of the door. Approaching a sliding door for access requires only a sideways push action as opposed to the two actions required for

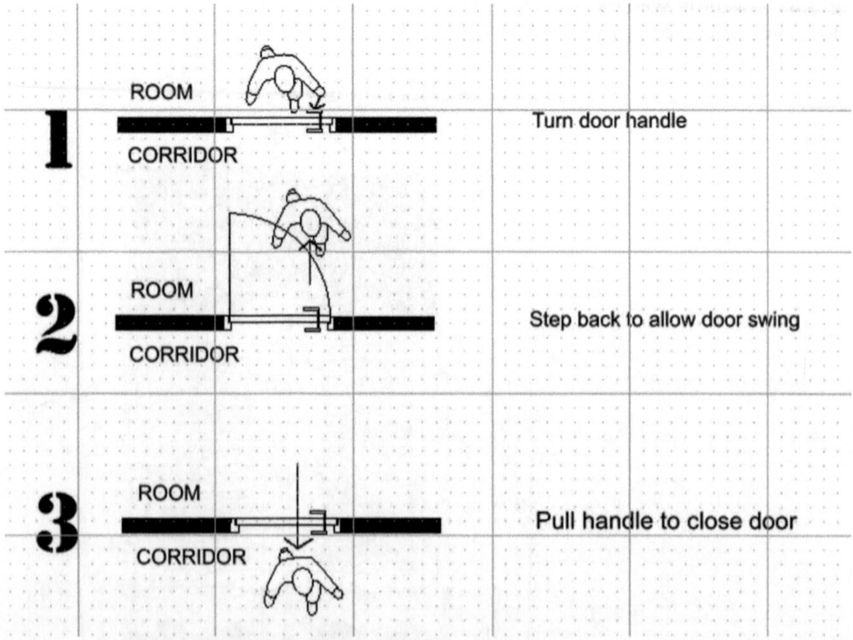

Fig. 5.64 Three stages to opening and closing a hinged door

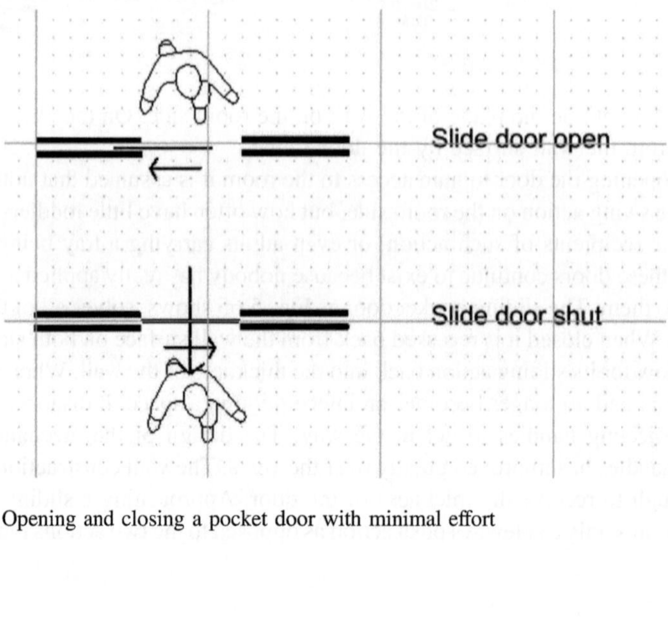

Fig. 5.65 Opening and closing a pocket door with minimal effort

Fig. 5.66 Sliding pocket door in Villa N1 in Sweden designed by Jonas Lindvall. *Photograph* Åke E: son Lindman

a hinged door of turn and forward push. The angry door slammer will have to find some other way of venting off steam. The sliding pocket door may take a little getting used to but if I have stated my case right it should be a massive improvement.

Figure 5.67 shows plans of a four-bed-terraced house in a proposed new estate development in St Albans, UK. This is the first time I have seen sliding pocket doors being used which is gratifying that my own proposal is approaching acceptability. It is also interesting to note that two living rooms are provided which suggests that a family who maybe divided in their choice of activity can pursue these without conflict.

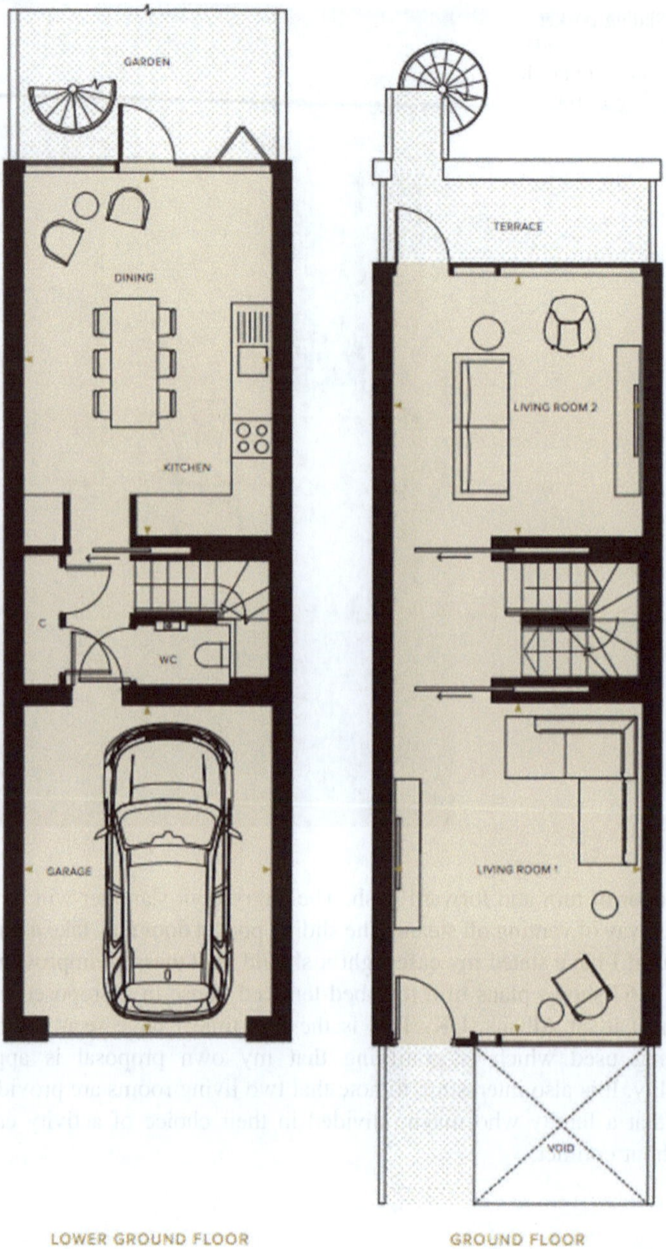

Fig. 5.67 Plans showing sliding pocket doors adjacent to the stairs. *Courtesy* Architects Benson and Forsyth

5.7 Conclusion

In my research for this book I have found great resistance to my proposals from the major mass housebuilders in this country because they are constrained by government targets and current space standards, as well as seeking high profit margins. The smaller builders who make fewer bespoke houses that are of high quality could not compete on price. A common point of opposition to my proposal is that it would cost too much compared with the cheap flat pack boxes that are made. To which I reply that my project is in a very different ballpark, which needs time and investment to massage it into the realm of realism. Costs do not form part of the initial impetus of this idea. This idea is like a seed that needs nurturing if it appears to be acceptable and worth pursuing. The urgent need for more housing to be built whilst conserving energy and achieving sustainability targets, has made the factory production processes more appealing in terms of their reduced cost, rapid supply and erection methods. In my view they do not have to be just used for social housing. The greatest impediment to such solutions is the cost of land. My next chapter explores various production processes and materials with the help of other collaborators.

Chapter 6
Building Technology

The strength of the prefabricated house lies in its popularity, its cheapness and the industrial base from which it operates. These are precisely the areas in which modern architecture is weakest. Modern architecture is unpopular, expensive and divorced from industrial production.
The government thinks that housebuilders should invest in factory production and begin making reliable, high quality houses that are cheap to run as well as cheap to build. Colin Davies (*The Prefabricated Home*, Colin Davies. London, Reaktion Books, 2005)

Abstract My aim is to investigate the most appropriate means of prefabrication to eco-standards that will enable my house design to be made into a reality. I have openly relied upon a variety of sources varying from journals, to specialist professionals, and organisations connected with this field of work. I summarise the choices of construction and building services as well as focusing upon advised solutions for my own house design as seen in Sect. 5.4. In that section, I recommend metal roller shutters over all windows and doors as a security measure. It amazes me that, whilst they have always been popular in Germany, they have never caught on in the UK.

6.1 Prefabrication

Building with prefabricated systems encompasses the production and use of pre-planned components or modules as a solution to build with higher quality and more efficiency. It is associated with dimensional grids, high technical standards, lower costs and the repetition of components or objects. Prefabricated Systems. By Knaack, Chung-Klatte and Hasselbach. Published by 4

Prefabricated systems emerged after WW2 as a rapid means of housing the displaced and homeless. It was a different matter of urgency as compared with the present crisis, due to the destruction of buildings in our towns and cities. The problem is that today's factory produced housing does not necessarily mean that the quality of provision is being improved. Meeting the demand means building

A. Sully, *The Estate House Re-designed*,
https://doi.org/10.1007/978-3-319-90397-2_6

at economic rates but building thousands of dull boxes, in my opinion, does not serve the public well. It is astonishing that in America only 9% of all housing starts are manufactured or factory built. This is because there is an unfortunate stigma associated with prefabrication as though it lacks quality and is mainly for social housing. It does not help to find that the design of this sector of housing in the USA is lacking with very dull boxes produced and poorly planned.

The long-term failure of successive Governments to build enough housing to meet growing need is widely accepted. In the 12 months to June 2015 there were 131,060 housing completions in England; this compares with estimates of a requirement for housing in excess of 230,000 units a year. Within this context, there is a view that increased use of off-site construction methods can help to meet housing demand. It is (relatively speaking) easier to deliver homes to higher quality standards using factory made and pre-assembled products than it is using traditional construction techniques. In addition to reduced costs, off-site construction can offer higher quality finishes, cheaper construction financing and reduced working capital requirements.

Mark Prisk said:

Off-site construction is an industry with untapped potential in this country, and could in time revolutionise the way we deliver our housing, providing a swift, high quality solution to creating cost effective, zero carbon homes.

House of Commons Debate, November 2015

'Onsite' (as opposed to 'Off-site' above) refers to modern methods of construction, which brings together systems or components which are predominantly assembled on site. One has to remember that there is a distinction between factory-produced housing intended for mass production and smaller housing projects that use a prefabricated building system with new hi-tech components. My research reveals that there are very few factories capable of meeting current demand, which is why the Chinese company China National Building Material Company is investing in a factory: A UK housing association has signed a landmark £2.5 billion joint venture with a Chinese state-owned construction company to build 25,000 modular homes over the next five years. I contacted the architects and they confirmed that they will be producing boxes, albeit with sound construction and to Eco Home standard (such as Passivhaus see Sect. 6.7). The Berkeley Group are also building factory-built homes in Greenwich:

The Properties are built as modules in a factory in the Midlands with each room being packed in a steel box frame, transported to the site and then connected together. The system saves approximately 19 weeks compared to traditional on site bricks and mortar building methods and with closer quality control cuts down on building defects. www.Property118.com

Prefabrication, or off-site fabrication, refers to parts of a building, interior or exterior, that are produced and assembled in a place other than the building site. Deborah Schneiderman[1]

[1]*Inside Prefab*, Deborah Schneiderman. New York, Princeton Architectural Press, 2012.

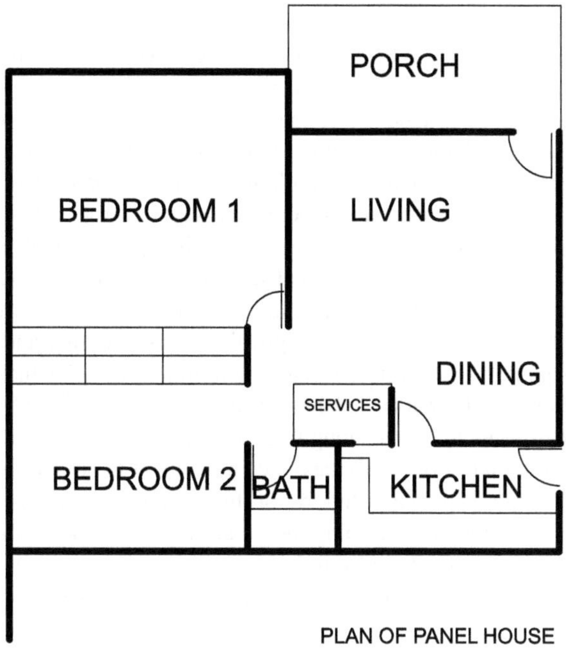

Fig. 6.1 'The Packaged House': plan of mass-produced panel house designed by Gropius and Wachsmann. 1940. *Drawn* Author

Prefabrication is a technique used in a variety of industries, where components are manufactured in a factory, which may be assembled into a finished unit and transported as a total package, or may be dispatched as parts for final assembly on site. The prefabrication process is particularly useful in the construction when a fast reliable solution is needed. Richard Abbott of Aganto Ltd. 2014

There was a book published in 1984 by Gilbert Herbert entitled 'The Dream of the Factory-made House' which told the story of the collaborative effort in 1940 between Walter Gropius,[2] then at the height of his fame, and Konrad Wachsmann,[3] both in exile in the USA from their native Germany (Fig. 6.1).

Due to WW2, there was a demand for houses and other buildings that could be built quickly, so the idea of factory-made houses was a logical path to follow. Unfortunately, the dream did not come true despite setting up a factory that made 200 houses but only a few were sold. The module they chose was about 1 m, which is not far off my choice of 950 mm. Advancements made in technology and industrial fabrication during the war share a symbiotic relationship with a systems approach in architecture.

[2]Walter Adolph Georg Gropius (1883–1969) was a German architect and founder of the Bauhaus School, who, along with Ludwig Mies van der Rohe, Le Corbusier and Frank Lloyd Wright, is widely regarded as one of the pioneering masters of modernist architecture.

[3]Konrad Wachsmann (1901–1980) was a German modernist architect although he trained as a cabinetmaker. He is notable for his contribution to the mass production of building components.

National Homes of Lafayette, Indiana, USA, started producing factory-made homes in 1940 and went on to become a very successful company turning out 750 homes a month based upon a prefabricated timber panel system. These were not designed by an architect but rather by engineers who had the skill, capacity and funds to make them. This highlighted a dilemma that faced architects in that for the first time their sovereignty or authorship of a building was being challenged. Their position up to now was to produce designs and specifications that instructed builders to build on a site through collaboration with engineers and other professional and trade specialists. Factory production methods took that seniority away from the architect and placed in the hands of the factory led engineering team. This echoes the time of the medieval castles and churches which were 'designed' by stonemasons and other craftsmen, artists and sculptors who were educated men and would eventually rise to the rank of master builder. The makers led the way and it was not until the fifteenth-century engineer/artist Brunelleschi[4] came along was the term 'architect' ascribed to anyone. He was one of the founders of the Renaissance movement in architecture and the arts. Over the next few hundred years, students of this area of work worked under the established architects of their day learning their craft, and it was not until the Royal Institute of British Architects was founded in 1839 was the profession given a home base from which over the next few decades approved schools of architecture were established.

6.2 Other Early Prefabricated Homes

'... less than 160,000 homes were constructed under the (postwar) programme, but the idea of a house that could be mass-produced in a factory caught the imagination of the public, and today we talk about prefabs as a generic label rather than 'temps' (for temporary homes) or the trade names chosen by manufacturers—Arcons, Spooners, Phoenixe and Tarrans.'

Palaces for the people/prefabs in post-war Britain—by Greg Stevenson. Published by Batsford 2003

There was an urgent demand for housing in countries affected by this war which standard building methods could not satisfy (see Sect. 1.6). Examples of prototypes that never made it to the factory production run but served as inspirational designs for students of architecture are as follows:

6.2.1 The Maison Citrohan—Designed by Le Corbusier

A machine for living in the house consisted of three overlapping floors. The ground floor houses the living room and dining room in front and kitchen and service areas

[4]Filippo Brunelleschi (1377–1446) was an Italian designer and a key figure in architecture. He was recognised to be the first modern engineer, planner and sole construction supervisor. He was one of the founding fathers of the Renaissance.

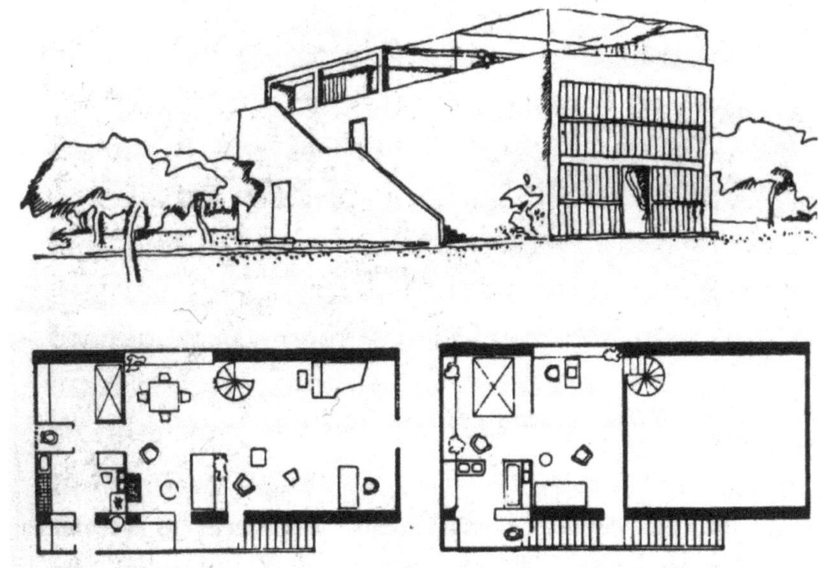

Fig. 6.2 The Maison Citrohan—designed by Le Corbusier 1922. Re-drawn by author

to the rear. The first floor was the master bedroom and accompanying bathroom, to be an intimate balcony space above the space of the two-storey living room. The second floor contained two individual bedrooms and minimal bathrooms in the rear and the garden terrace at the front, which had the possibility of reaching the roof. The structure was concrete frame skeletons with membrane walls 3 in. thick with cement projected on deployed palastro, leaving a vacuum of 20 cm (Fig. 6.2).

6.2.2 *Wichita House a Round House Designed by Buckminster Fuller*

The single volume interior contained a living/dining room, a small kitchen and laundry unit, and two bedrooms, each with their own bathroom and freestanding storage unit. It was suspended and supported by a central mast, from which a light shell enclosed living space was hung in tension. All utilities (heating, refrigerating, air conditioning, laundering, dishwashing, bathroom and kitchen units) were grouped around the central mast. The walls and roof were constructed of unfinished aluminium panels and the ribbon-like windows were made of plastic. Capping the building was a wide, aerodynamic sheet metal ventilator, designed to rotate and lift depending on the weather conditions. The ventilator drew in outside air, allowing the windows to be inoperable (Fig. 6.3).

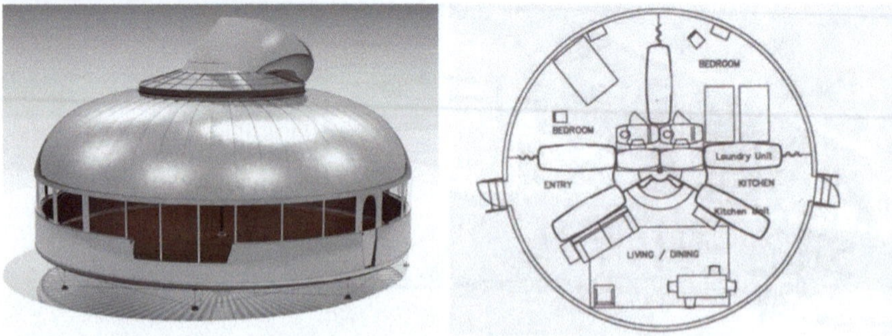

Fig. 6.3 Exterior model of Wichita house with plan showing services pods. From Tumbir.com

6.2.3 Early 1950s—Meudon House by Jean Prouvé

Only a few of these prefabricated houses were built in the Paris suburb of Meudon again not fulfilling the desire for mass production. Jean Prouvé was not strictly an architect or a designer or an engineer, but all of those things at once. Having trained as a metalworker, he set up his own studio in 1924. Until 1952 (the year when his factor was bought up), he worked there producing chairs, furniture, aluminium structures, modular houses and portholes. After this, he continued to practise his trade as a designer. The example of the 'standard house' developed in response to the post-war housing shortage was based on modules 1 m wide. It was designed to be assembled in 48 h. Ultimately, it produced very dull boxes and was an example of technology winning over human needs (Fig. 6.4).

The tension between the disciplines of architecture and construction was constant until the Hong Kong and Shanghai Bank was built. The architect Norman Foster, who always had a close relationship with the aircraft industry and its factory methods of manufacture understood that a 'product design' approach could open up huge architectural possibilities in the design of a building. It extended the palette of

Fig. 6.4 Meudon House, Meudon, Paris. Designed by Jean Prouvé. Early 1950s

Fig. 6.5 Hong Kong and Shanghai Bank by Foster & Associates 1986, courtesy the architects and the photographer Ian Lambot

choice and conception of how to approach the design of structure and all the components of building, which had been traditionally in the hands of manufacturers. This meant that a review of the structure was beyond the simple solution to an expressive suspension structure as described below. The Foster office called this 'Design Development' by working side by side with the manufacturer. The old team of architect, engineer, builder and shopfitter was replaced by an architecturally led factory production team for the first time ever (Fig. 6.5).

Quote from the architects:

The requirement to build in excess of a million square feet in a short timescale suggested a high degree of prefabrication, including factory-finished modules, while the need to build downwards and upwards simultaneously led to the adoption of a suspension structure, with pairs of steel masts arranged in three bays. As a result, the building form is articulated in a stepped profile of three individual towers, respectively twenty-nine, thirty-six and forty-four storeys high, which create floors of varying width and depth and allow for garden terraces. The mast structure allowed another radical move, pushing the service cores to the perimeter to create deep-plan floors around a ten-storey atrium. A mirrored 'sunscoop' reflects sunlight down through the atrium to the floor of a public plaza below—a sheltered space, which at weekends has become a lively picnic spot. From the plaza, escalators rise up through the glass underbelly to the banking hall, which was conceived as a 'shop window for banking'.

6.3 Modular Co-ordination

Whilst manufacturing techniques were improving throughout the building industry to assist with the prefabricated elements of housing, a controlling square grid was established of modules that suited certain materials as well as aiding the planning and design of building elements. This was called 'modular co-ordination' and created a series of a standardised system of products that were interchangeable and conformed to factory production methods. Schools and hospitals in Britain were some of the first type of buildings to use modular co-ordination. After many trials in the mid 1960s, a 3 ft 4 in. (almost a metre) module was chosen and Hertfordshire County Council was one of the first local authorities to apply this to their school building programme. Industrialised building systems grew out of the world of prefabrication to such an extent that building elements such as walls would be cast in concrete panels. It followed that cladding systems were developed improving the speed of erection and building on site.

However, modular co-ordination was not to last in the way it was implemented. Breakdowns of communication occurred between architects and factories according to the Buildings Economics Research Unit, and dimensional co-ordination proved to be more tricky in searching for accuracy on site. Another factor was that because of the monotony of the square grid, a rather dull architecture was produced. One building system that did survive well into the twenty-first century was Consortium of Local Authorities Special Programme (CLASP) mainly due to good co-ordination between all parties concerned.

6.4 Structural Options

Industrialised building is based on the principle that as much of the work as possible is transferred from the site to the factory leaving only a simple assembly operation to be carried out on site. There are two generic forms of industrialised building, namely 'closed' and 'open'. With closed construction the bulk of the structure is produced from a fixed set of pre-fabricated parts, allowing very little scope to substitute products from other manufacturers. Open building on the other hand produces a shell from a relatively small number of parts, thus allowing the designer considerable scope to create a unique design. Many Industrialised Building systems employ the large panel method of construction comprising factory-made pre-cast concrete floor and wall panels. These units arrive on site in their assembly sequence and are assembled with the aid of a crane. Non-Traditional Housing in the UK. By Keith Ross of the Building Research Establishment (BRE). 2002

The biggest issue facing house designers and builders in the modern age is how to ensure the construction accords with the worldwide need to exist within limits of sustainability. This involves ensuring that all resources used are made from renewable resources, instead of eating away at the eventually limited worldwide supply of the fossil fuels that we use to create gas and electricity, and that the emissions from the building can be minimised, and to aim for a zero emission target, or a non-carbon contribution. James Murray-White. Sustainablebuild.co.uk

The following Sects. 6.4.1–6.4.9 is by courtesy of The National Self-Build and Renovation Centre www.nsbrc.co.uk and http://www.the-self-build-guide.co.uk. The 'open' and 'closed' systems referred to in the above BRE paper are also described below.

6.4.1 Timber Frame

Timber frame construction refers to a range of systems from traditional post and beam to stick built, open and closed panel systems. Timber Frame accounts for around 70% of all new house builds around the world … and the NHBC find less problems with timber frame homes than homes built with masonry.

Open panel timber frames comprise a softwood stud frame backed with a layer of timber sheet material. The panels are manufactured in accordance with the house design, delivered to site and fixed together to provide the structural frame of the house.

Closed panels are similar in construction to open panels but factory prefabrication goes one step further with insulation and vapour barriers incorporated and plasterboard fixed to 'close' the panel. Service conduits, socket points, windows and doors can all be factory fitted further reducing the portion of on-site construction.

Timber is also a renewable resource; with the appropriate insulation between the studs and in the panels, the energy performance will be very high. A growing number of frame suppliers are supplying engineered structural beams (Glulam for example), which enable a greater span to be achieved, offering better flexibility for ground-floor room layouts. Cavity walls were introduced about 100 years ago to control the ingress of damp through solid external walls. The external cladding on the outside, which can be any material including masonry, timber or render, is only for weathering and decorative purposes, rather than taking any of the loading from the roof.

Prefabricated panels manufactured in accordance with the house design are delivered to site and erected forming the inner skin of the house—similar to the blockwork skin in masonry construction. This is the structural load-bearing element of the house and supports floors, the roof and internal loads.

6.4.2 Oak Frame

This is another variation on timber frame, using 'oak post and beam' construction, with centuries-old jointing techniques. The oak can be left on display which gives final house an authentic period look. Oak requires very little maintenance, is very strong, and can last for hundreds of years. As one of the most natural of resources, its carbon footprint is very low when all elements are being considered.

The in-fill panels between the posts must meet stringent energy regulations and cope with the shrinkage of the oak. Green oak construction uses unseasoned oak with a moisture content of 20–25%. As it dries out, it shrinks across the grain. This causes gaps around the in-fill panel, which, with many systems, will at first need to be filled regularly. Waterproof expanding foam and a sliding joint should be incorporated to keep the effects of movement to a minimum.

6.4.3 Steel Frame

Another option is steel frame construction, where the timber studs are replaced with metal. As with timber frame, self-builders have a complete choice of external facing material, provided that it is properly connected to the steel studs which take the entire load from the roof. Steel frame is often considered where large clear spans are required. Steel frame construction is not particularly common in the UK. A number of manufacturers have attempted to supply steel systems but, due to the exacting standards, they are not always perceived to be as flexible as other forms of construction. The steel framework is erected on site, very much as large industrial units are built. You are required to wrap the whole exterior of the frame with a rigid insulation layer, to prevent heat loss through cold bridging of each stud to the cavity. Additional insulation will then be placed between the studs.

High energy efficiency is possible with steel frame, and good acoustic standards can be achieved. Steel frame is still not common in the UK; its use has increased significantly in the USA over the last five years and is expected to increase over here too. Few companies will deal with one-off houses; therefore, prices are difficult to gauge, and however, you should expect costs to be in the order £1000–£1200 per m^2.

6.4.4 *Insulated Concrete Formwork (ICF)*

This is an ingenious system where the structure of the house is built out of hollow polystyrene blocks known as insulated concrete formwork (ICF), which are then filled with concrete. Externally, there a number of choices including a brick skin, an external render directly onto the polystyrene, or a reconstituted stone slip system glued to the face of the polystyrene.

This system is ideal for those who want to have a concrete block and beam system as their intermediate floor, lined with a battening system or appropriate clips and a plasterboard ceiling. This type of solid floor is particularly suited to underfloor heating systems and also offers very good acoustic insulation; however, service routes within the intermediate floor will need careful planning. The assembly of the hollow polystyrene blocks is reasonably quick; once assembled, the internal cavity of the blocks is filled with ready-mix concrete which provides strength. The result is a concrete external wall, wrapped on both sides with a good layer of polystyrene insulation. Care must be taken not to leave any voids in the concrete, particularly around structural openings; to this end, the ready-mix concrete is vibrated within the cavity using a compressed air poker. It is equally important not to get too ambitious with the pour—too much concrete will cause the blocks to burst! There are some important techniques involved in this method, including the metal ties which hold the polystyrene structure together before the concrete is poured in, and some external bracing which can be fundamental to the erection. Lintels over openings are formed in the same way, with steel rods over openings to an engineer's design. On the inside, there is no need for additional insulation, so plasterboard can be adhered directly to the polystyrene.

In many respects, insulated formwork is ideal for the self-builder because the structure and wall layout are relatively quick, easy and—being polystyrene—not difficult to lift. It is not yet considered a mainstream method (it is probably used by only 1–2% of self-builders), but it has certainly become more popular, with more suppliers coming to the market in recent years. A costing of £1000–£1200 per m^2 would be a fair estimate, but this could be reduced by carrying out a lot of the build yourself—the system is fairly straightforward so this should be within most self-builders' capabilities although building tolerances can be difficult. ICF also offers a relatively short build time and currently meets all building regulations.

6.4.5 *Brick and Block*

The brick outer skin is indeed the façade of a very 'traditional construction' method, commonly referred to as 'brick and block'. The external wall has a blockwork inner skin which is connected to the external brickwork using wall ties. Lightweight blocks are an efficient way of building, being easy to work with so that a relatively large volume of work can be undertaken in a short period of time. Blocks are manufactured to take different loadings and it is important to understand that higher strength blocks are used in certain scenarios. The blocks are light because of the air trapped within them, which keeps the warmth inside the building making them good insulators. A normal size of single block would not meet energy regulations on its own though, so additional insulation is usually inserted into the cavity or packed on the inside of the blockwork by using a composite thermal board or vertical studs with insulation between. Packing out on the inside in this way is also very helpful for electrical and plumbing service installations. The blockwork inner skin is the load-bearing element with this system, and although the brickwork is generally built at the same time, to a large degree the brickwork is weatherproofing again; it is tied to the blockwork usually with stainless steel ties to add structural rigidity to the masonry panels. Bad weather in winter can affect your programme as construction does have to stop in adverse conditions such as heavy rain and frost. Brick and block is the most popular choice for self-builders in the south of the UK, where weather delays are less likely. Most people perceive a brick and block house to be more 'solid' than timber frame and are therefore prepared to pay a premium. It is much more difficult to reach the U-values and airtightness required under current regulations and these requirements become more stringent all the time. The 'normal' insulating block is 100 mm thick, but to get improved levels of insulation you may want to consider using a 150-mm block, although this increase should be primarily for structural reasons, and possibly upgrade the cavity insulation—of course, all of this adds to the cost. With a 100-mm cavity, fully filled with the top-of-the-range DryTherm 32 insulation, and a structure of brick, cavity, insulating blockwork and dry lining, we achieve around 0.25 W/m^2 K. Generally, you should budget around £900–£1100 per m^2.

6.4.6 Thin-Joint System

Thin-joint system is a variation on a traditional brick and block construction. The blocks used are aerated blocks (non-aerated can do as well) offering good thermal qualities and are laid using a cement-based adhesive rather than sand and cement, so the joint between blocks is very thin. This enables very rapid construction, and more blockwork can be undertaken per day. Because there is more blockwork and less jointing material, it also dries very quickly. The thin joints also offer far better airtightness which assists the thermal qualities, than a traditional 10-mm mortar bed. The internal skin of blockwork comes in a variety of thicknesses and is improved thermally by adding insulation, either in the cavity or on the inside, with composite thermal plasterboard. With certain thicknesses, however, no additional insulation is necessary and these blocks can also be used for solid wall construction using waterproof renders on the outside. This method will cost in the region of £1000–£1200 m^2.

6.4.7 Pre-cast Concrete Panels

Reinforced pre-cast concrete panel construction system provides an exceptionally strong building, would be good for higher rise construction and are also ideal for basement construction as long as a suitable waterproofing, (tanking) system is used. The intermediate floor and the panels are fixed with special metal panel pins and dowels and bolted where necessary. Dense concrete like this is a good conductor of heat so the system on its own has practically no insulation qualities; therefore, you have to insulate in the cavity on one side before the blockwork and render is fixed, and on the other side only on the interior, because the render is fixed directly to the concrete panel. When insulating in the cavity, you effectively trap the concrete on the inside of the thermal layer, and in these situations, the concrete itself will become highly efficient for thermal mass storage.

Pre-cast concrete is relatively uncommon for the construction of a one-off house (although ideally suited for basements)—the system lends itself far better to blocks of flats or a terrace of houses on larger site—heavy plant and plenty of space is needed to put the panels in place and the panels are made in a formwork mould. For an individual house, you should budget £1200–£1500 per m^2, but be prepared to be quoted higher.

6.4.8 Structural Insulated Panel System (SIPS)

A traditional cavity wall construction whereby the inner skin is a structurally insulated panel (SIP), and the external facing is largely there for weathering and decorative purposes. This is a similar system to the more common 'open panel' timber frame system, but in this instance, a fully insulated and completed panel, often pre-fitted with windows and doors, will be manufactured and delivered to site. The panels themselves are more expensive to make, but savings are introduced later in the build as you are unlikely to need further insulation in the walls to meet current standards. The panels carry the building's structural load, similar to other timber construction methods. The panels are faced on both sides with orientated strand board (OSB), and the two sides are factory bonded to a rigid insulation, to make a wafer-like construction. The principle is that very large panels can be constructed, using much less conventional timber, as studs are only required at junctions. The larger panels are craned into position so good access to site is essential, but this does make for a very short on-site construction time. A number of the manufacturers will also fix timber battens on the inside face of the SIPS, to ease the installation of pipework and cabling; this also provides good fixings for the plasterboard. Internal walls can be open studwork if insulation is not an issue. Using SIPS is probably one of the quickest methods available for building and is very thermally efficient. However, it is not so easy to adjust on-site if changes do need to be made. Using SIPS will give you an overall costing of £1000–£1200 per m^2. Less than 5% of self-builders currently use this system, although there has been a growing interest in recent years.

One such by-product is the award-winning, British-made and patented Mantle Building System which is an hybrid solution between Stressed Skin Panels and regular Structural Insulated Panels (SIP) systems. The outer layers of the panel are cement particle board. The inner core of the panel is a special form of expanded polystyrene, manufactured to less than ±1% tolerance to give its consistency of strength and insulation. Their combination produces a composite system at least 4 times stronger than engineered timber solutions. The standard panel has a thermal insulation U Value of 0.10 (before cladding and finish), an acoustic insulation starting at 60 dB and an airtightness of about $1 m^3/m^2$@50 Pa. Its structures are about 1/15th the weight of traditional build and it has less than 30% of the embedded carbon of its nearest competitor, timber. There are just three basic system components:

Panels (2.4 m × 1.2 m × 220 mm)
Ring Beams (2.4 m × 350 mm × 198 mm) to hang floor joists, if it is necessary
Corners (2.4 m × 500 mm × 500 mm)

It has been used across the entire build spectrum: housing, multi-storey apartments, schools, offices, sports complexes, care facilities, etc.

From Sip Building Systems:

The standard SBS Structural Insulated Panel consists of two facings of 12-mm OSB 3 (oriented strand board British Standard BS EN 300 for structural integrity) bonded by pressure injection to CFC free/OPD zero polyurethane closed cell foam. The facings and core act as a composite structure, and the option of various facing thicknesses is available upon request.

Dimensions: Thickness from 75 to 250 mm. Standard width 1220 mm.
Insulation: u-values can range from 0.35 W/m^2 K down to 0.10 W/m^2 K.

Acoustic performance can be improved by the interior lining of plasterboard.

Durability: A life of 60 years can be expected.
Air tightness: A good level can be achieved but a mechanical ventilation system is also recommended.
Structural loading: Using a factor of safety of 2.0, it is seen that panels will comfortably support a vertical load of 90 kN/m. These loads are well in excess of the average foundation loading from a typical two-storey house of traditional construction.

Pests: The rigid urethane insulation core and facings used in the manufacture of SIPs resist attack by mould and microbial growth, and do not provide any food value to vermin.

Fire: SIPs exceed the Building Regulations and British Standards requirements with regards to fire resistance. Tests suggest they can achieve up to 73 min resistance.

Service Runs: SIPs companies prefer services to run through internal partition walls. Wiring is run through the cavity between the panel and the internal plasterboard or, preferably, chased out through a double layer of plasterboard (which removes the need for a cavity and creates a more solid feel and sound to the wall, reducing noise issues). Wiring should be protected with an IEE-approved metal guard.

6.4.9 *Cross-Laminated Timber—CTC*

Cross-laminated timber (CLT), also known as X-LAM, is a structural two-way spanning, prefabricated panel used to form environmentally sustainable, structural walls, roofs and floors across a wide range of structural applications in the residential, education, commercial and sports and leisure sectors. Thanks to the gluing of both longitudinal and transverse layers which reduces the working behaviour of the wood to a negligible degree—CLT meets the standards required by modern building materials.

B & K Structures brings to the UK construction market a seamless and consistent supply of cross-laminated timber whilst offering an unrivalled level of information and technical data to all professionals at every stage of the planning, design and construction process. B & K Structures can offer both pure CLT systems and the possibility to incorporate cross-laminated timber into a hybrid structure ensuring the best possible sustainable structural solution for each individual project.

In partnership with Austrian manufacturer Binderholz, B & K Structures formed the X-LAM Alliance which is at the forefront of the European research and development in the cross-laminated timber sector. The X-LAM Alliance brings together experts in design, engineering and construction to deliver a robust, reliable and sustainable solution.

6.5 Specific Building Elements not Covered by Above

6.5.1 *Posi-Floor Joists*

With the move towards the building of more sustainable homes, an increasing number of mechanical ventilation and heat recovery (MVHR) systems are being installed in the UK and Ireland, driving more and more architects and specifiers to look for a versatile open web flooring solution. This system combines the lightness of timber with the strength of the Posi-Strut™ steel web, enabling longer spans than would be possible with alternative timber products, allowing a variety of internal room layouts within an external shell (Fig. 6.6).

Fig. 6.6 Mitek posi-joist, courtesy Mitek

6.5.2 *Structural Vacuum Panel*

THE PROBLEM: We wanted to build passive modular buildings, but with a maximum module width of 4 m we could not have external walls that were 400 mm thick so we needed to design our own 'thin-wall' super insulation system.
THE SOLUTION: We have developed the 'Structural Vacuum Panel' (SVP) and this is an engineered hybrid timber panel that integrates vacuum insulation technology and can be used to construct walls, roofs and floors. Depending on the application, the thinnest SVP is just 37 mm thick and achieves a *u*-value of 0.16 W/m^2 K. Thicker panels include additional PIR, a 125-mm SVP has a *u*-value of 0.09 W/m^2 K. Similar in appearance to a standard closed panel or SIP, however thermally its far superior. If you are designing a passive house, then every square millimetre of usable floor area counts. By using the SVP, you could increase the usable area by up to 15%.
PIR boards (Polyisocyanurate—a thermoset plastic) are the most popular type of rigid insulation **boards** currently available. They are closed-cell, rigid foam **board** insulation, that are primarily used on the roofs, walls and floors of domestic housing, offices, retail and industrial warehousing facilities and educational premises, to name but a few.

6.5.3 *Makers of Pocket Doors (see Sect. 5.6)*

The pocket door is a means of sliding the door away into the thickness of the adjacent wall, thus making its presence disappear and the door opening is left clear of any obstruction. A variety of door designs are available including part glass if required.

The doors can be either top hung or gliding on a floor track, or a combination of both. A recessed lockable handle is appropriate as can be seen in the illustrations of one by Hafele and one by Halliday and Baillie (Figs. 6.7, 6.8, 6.9, 6.10, and 6.11).

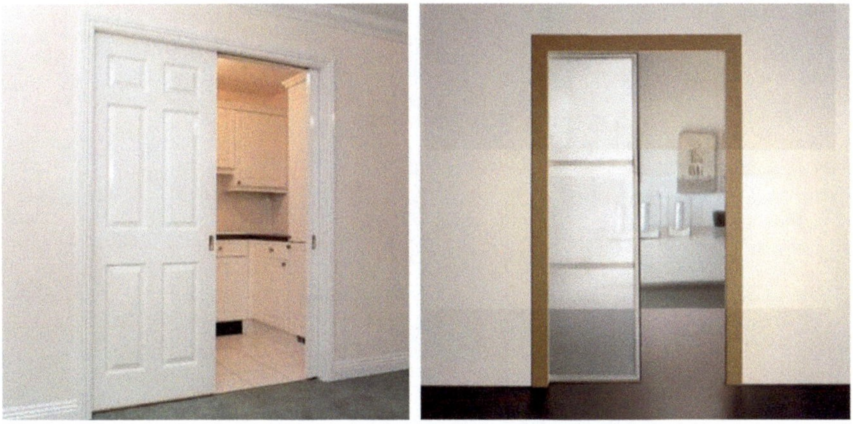

Fig. 6.7 Pocket doors courtesy Hafele UK. OPTIMO on right

Fig. 6.8 Recessed slide handles from Halliday + Baillie HB695

Fig. 6.9 Handle from Eclisse Syntesis Collection

Fig. 6.10 Eclisse Syntesis sliding pocket door

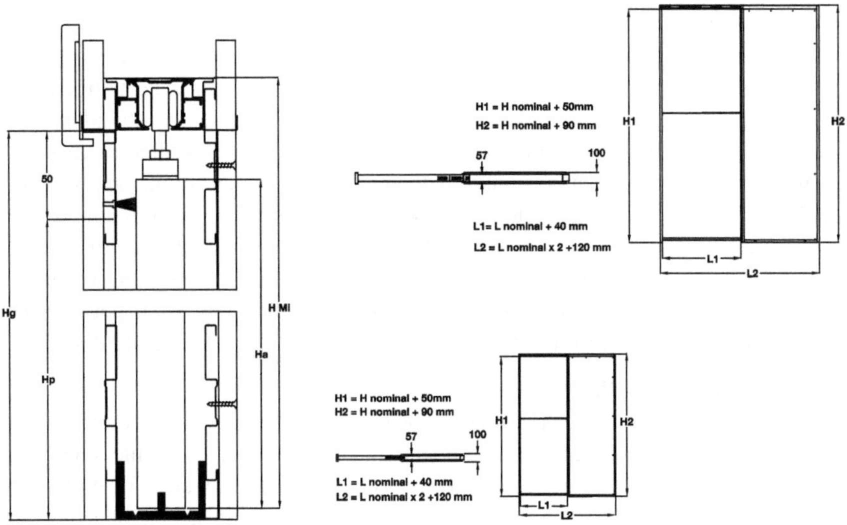

Fig. 6.11 Detail drawings of the Hafele OPTIMO sliding door floor glide and overhead track

Eclisse Syntesis Collection. The pocket door system without jambs or architraves. Integrated into the wall, a simpler finish with no extraneous lines.

6.6 Building Services

6.6.1 Heating and Hotwater

Solar Heating
Solar thermal system just supplies hot water.

Solar Photovoltaic (PV) system: Solar PV panels require only daylight, not necessarily sunlight to operate; therefore, they are capable of generating electricity even on cloudy days. Solar PV will work equally well in urban or rural locations and can be installed on the roof of most homes that can support the equipment. They can easily be connected to a building's electricity supply and any surplus energy can be sold back to the National Grid. When in use, Solar PV do not generate greenhouse gases or cause pollution. Each kilowatt peak (kWp) of electricity produced can save approximately 455 kg of carbon dioxide emissions compared with electricity generated from fossil fuels.

Biomass—Example: the wood pellet stove. Introduced from the Nordic countries and still in its infancy in this country. Useful backup to solar heating

Air Source Heat Pump
Even cold air is full of energy and air source heat pumps use the freely available heat in the ambient air to provide efficient heating and hot water at air temperatures as low as −25 °C. Because the source of the heat—the air—is abundantly all around us, air source heat pumps have the advantage of low installation costs and minimal space requirements, making them ideal for new-build applications. Suitable for either indoor or outdoor installation, air source heat pumps can be used for heating, cooling and to produce domestic hot water. With the relatively mild winter temperatures in the UK, air source achieves seasonal coefficient of performance comparable with ground source, meaning excellent levels of performance can be achieved throughout the year for a quarter of the price as ground source.

Ground Source Heat Pump
The earth stores an enormous amount of solar energy both from solar radiation and rainfall, and ground source heat pumps extract this freely available energy to provide low cost, low carbon, high efficiency heating and hot water. To extract the energy, ground source heat pumps use collectors, filled with a mix of water and antifreeze, buried in the earth, either horizontally or vertically. Although more costly to install than an air source heat pump, with the constant temperatures below ground throughout the year, ground source heat pumps deliver high levels of efficiency all year around irrespective of the seasonal variations in air temperature.

6.6.2 Ventilation

The ventilation system can be used for night-time cooling. The low-watt fan actively ventilates the house and filters out pollen. Mechanical Ventilation Heat Recovery unit: With airtight new buildings from ECOHAUS Company, whole house ventilation is the most effective ventilation strategy on the market. With addition of heat recovery, it continuously extracts humid air from wet areas such as kitchen and bathrooms and, using a heat exchanger, transfers the heat to filtered incoming fresh outside air which is redistributed to the living areas and bedrooms. Heat exchangers are up to 93% efficient which significantly helps reduce heating bills, plus the effect of condensation is dramatically reduced.

6.6.3 Electrical

LED/low-energy lighting

Megaman and LightwaveRF: Over the past few years, Megaman has established themselves as a leading manufacturer of low-energy light bulbs. Megaman UK is the exclusive distributor in the UK for LightwaveRF—a range of intelligent dimmers, radiator valves, sockets and sensors that make homes a little bit smarter.

CAT 5 wiring throughout fort broadband connection facilities and satellite points.

6.6.4 Security

It appears that most housing developments have burglar alarm systems fitted. It will cost more to be connected to a police station via a monitoring contract. There are two types of monitoring contract: keyholder and police response. In either case, you pay a monthly or annual fee, and then the monitoring company will alert a nominated keyholder or the police when the alarm goes off.

The second measure to be taken is window roller metal shutters as I recommend in my house design in Sect. 5.1. The company UK Roller Shutters produce a wide choice of shutters for domestic and commercial use including the ALU-LNTL model (Fig. 6.12).

The ALU-LNTL model of roller shutter is made from an extruded aluminium matrix. The double-wall aluminium slats are particularly effective at repelling burglars. The ALU LNTL can also be zinc coated to BS EN 10131:2006 to combat any possible corrosion with a powder-coated finish to any colour adding yet another barrier making them excellent in saline conditions such as coastal locations.

Fig. 6.12 Example of window shutters with the control box being hidden in the structure of the building, courtesy UK Roller Shutters

6.7 What Is an Eco Home?

An Eco Home is an environmentally low-impact home designed and built using locally sourced materials if possible, and harnessing a technology that reduces its carbon footprint, lowers its energy needs, and reduces the building's negative impact on human health and the environment, through better design, construction, siting, operation and maintenance.

- Glass that has two or three layers with a vacuum in between to prevent heat loss
- Reduce energy consumption and waste
- High levels of internal and external insulation
- Better than normal airtightness
- Good levels of daylight
- Good double or triple-glazed windows
- Passive solar orientation—glazing oriented south for light and heat
- Thermal mass to absorb that solar heat
- Minimum north-facing glazing—to reduce heat loss
- Mechanical ventilation with heat recovery (MVHR) system
- Heating from a renewable source (such as solar, heat pump or biomass)
- Photovoltaic (PV) panels, small wind turbine or electricity from a 'green' supplier
- Natural materials—avoidance of PVCu and other plastics
- Rainwater harvesting.

MVHR—Mechanical ventilation and heat recovery **systems**
PVCu—Polyvinyl chloride unplasticised.

6.7.1 Energy Efficiency

Approximately 45% of the UK total carbon emissions are a result of energy consumption in buildings.

SAP calculations are a requirement of the Building Regulations and are required for all newly built dwellings in the UK. A SAP Rating has been required for all new homes under Part L of the building regulations since 1995; therefore, most developers will be familiar with it. SAP stands for 'Standard Assessment Procedure'. It is the only official, government approved system for assessing the energy rating for a new home. SAP assessors must be accredited and registered with a certification body.

6.7.2 Carbon-Positive Homes

A long-term sustainable solution to global warming and climate change requires that we eliminate or substantially reduce the amount of carbon being emitted into the atmosphere from human (anthropogenic) activities. It can be achieved by creating carbon zero or carbon positive homes.

6.7.3 *Passivhaus*

Passive buildings, as distinct from Passivhaus, rely on high levels of insulation and tight control over airtightness. Buildings built to this standard can run effectively on low volume heat recovery ventilation systems required to maintain air quality. The main definition of what Passivhaus is (Fig. 6.13):

> *A Passivhaus is a building, for which thermal comfort can be achieved solely by post-heating or post-cooling of the fresh air mass, which is required to achieve sufficient indoor air quality conditions—without the need for additional recirculation of air.*

Information About a Passive House (From www.greenmatch.co.uk)

What Is a Passive House?

A passive house is a concept that originated from Germany, where it is known as Passivhaus. It is important to note that a passive house is not a brand name, but a construction concept that can be applied by anyone—commercial buildings as well as residential houses. A passive house is built so that it is very energy-efficient and comfortable at the same time without using a traditional heating system or active cooling. The passive house uses passive heat sources—the energy of the sun, the heat from running household appliances and the extracted air—in order to cover most of the heating demand. The remaining energy needs in a passive house should be supplied by renewable energy sources, such as solar panels or heat pumps. The active utilisation of this existing energy is one of the main characteristics associated with a passive house. Passive houses optimise thermal gain, at the same time minimise thermal losses. By using existing temperature, the required heating energy of a passive house is 90% lower than the energy needed for a conventional home.

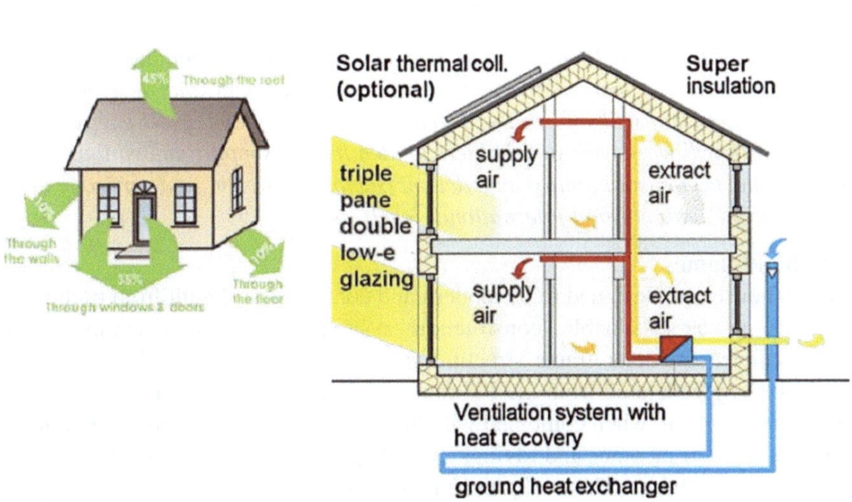

Fig. 6.13 Passivhaus diagram. *Courtesy* Passivhaus

Requirements for a Passive House:

There are certain specifications that a house needs to fulfil in order to be considered a passive house. There are three basic requirements for a passive house:

1. A passive house needs to have heating demands lower than 15 kWh per square meter annually (kWh/m^2 a).
2. A passive house needs to have primary energy demands (such as warm water, heating, house appliances) lower than 120 kWh/m^2 a.
3. A passive house must pass a pressure test and the pressure must be limited to 50 Pa with the pressure differential not exceeding 0.6 times a room's volume per hour ($n50 < 0.6\ h^{-1}$).

In order for a passive house to be able to have such low-energy demands, it needs to be built in a certain way. A passive house needs a high level of insulation, 'thermal bridge-free' construction (a thermal bridge is heat transferred through a poorly insulated part of a wall), airtight windows and doors, a mechanical ventilation system with heat recovery and heat efficiency, and it can make use of passive solar gains, although this is dependent upon a flexible design strategy. Moreover, a passive house should also have low-energy appliances and energy-efficient lights.

My research has discovered that sometimes designing a house to Passivhaus standards can be restrictive as far as the design is concerned, but this will depend upon how the design is accommodating the standards.

6.7.4 From BRE Innovation Park

BRE—British Research Establishment

Our vision for the BRE Innovation Parks is to establish a unique and unparalleled research and demonstration network, which inspires and develops innovative solutions that will inform the development of the global built environment. This network will work with stakeholders to identify and support innovations which demonstrate true sustainability and deliver social, economic and environmental benefits at the national and international levels.

Zero build homes

ZBHCo houses are designed to be modern and contemporary, built from traditional materials in a sustainable construction system. The system combines an energy-efficient modular frame structure with an integrated solar PV panel roof. The electricity generated from the PV roof panels earns revenues from the feed-in tariff scheme which, when combined with the free use of the electricity stored in the batteries, leads to incomes and savings exceeding the residual cost of electricity—a zero energy bills home. This will protect households against ever-rising energy bills and help to reduce fuel poverty.

Sustainability
Zero bills have maximised the use of UK-sourced materials to minimise the carbon footprint. The timber studs, OSB, and steel are all UK sourced. They maximise the use of natural building materials and eliminate urethane foams. This makes the building one of the lowest embodied carbon buildings possible. Combined with the BIPV solar roof, battery storage and zero fossil fuel energy systems, the home starts paying back its embodied carbon by generating more electricity than it needs over a typical year.

Heating, mechanical ventilation and hot water
This home is powered by a unique Domestic Hot Water System, Space Heating and MVHR Solution Developed by ZED Factory in conjunction with leading product manufacturers and utilising the passive solar design of the Zero Bills Home. Domestic hot water is provided to the tank via the heat pump compressor extracting heat from the black panel in the solar space. Warmed up refrigerant gas is passed through the compressor and exchanger providing heat to the tank in the conversion of the hot gas back to its cooler liquid state. DHW water takes priority over space heating. MVHR, the heating system, is used to meet the ventilation rates of the building and provide heat recovery at 90% efficiency.

Rainwater Harvesting

An integrated rainwater storage Bluroof system combined with green roof overflowing into rainwater.

6.8 Contribution from Hanse Haus

The following specification and construction methods for a typical prefabricated two-storey house (see Sect. 2.5.4) are taken from their literature.

Hanse Haus are a member of the Quality Association of German Prefabrication and their construction techniques achieve a high rating in accordance with the Code for Sustainable Homes.

6.8.1 Exterior Walls 310 mm Thick

How our exterior walls are constructed, from the outside in:

Approximately 2-mm synthetic resin render with a rough finish.
Choose from a selection of colours.
Approximately 3-mm reinforcing render with glass fibre tissue.
150-mm thermal insulation, WLG 032 8-mm OSB board, bonded to the timber frame across its whole surface to EN 300 standards.

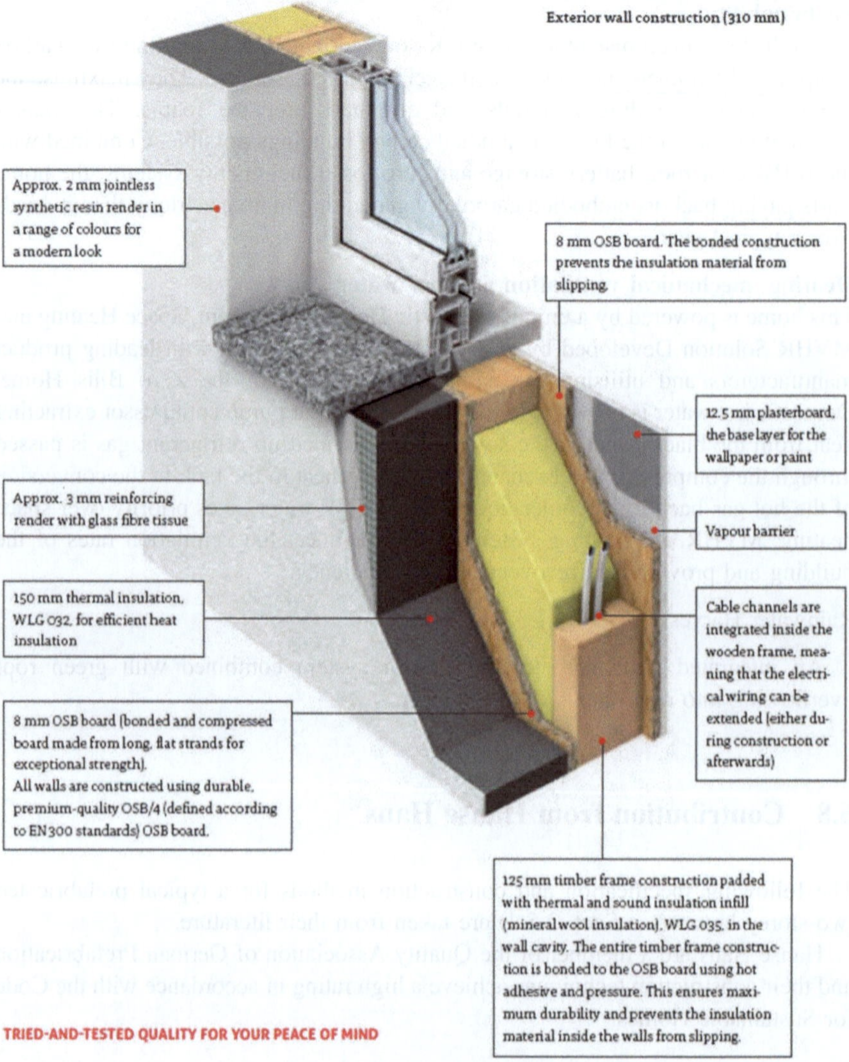

Fig. 6.14 Typical Hanse Haus external wall construction

125-mm timber frame construction padded with thermal and sound insulation infill (mineral wool insulation). WLG 035.
8-mm OSB board bonded to the timber frame across its whole surface. Vapour barrier.
12.5-mm plasterboard, screwed to the OSB board.
Heat transition coefficient in wall cavity only 0.116 W/m^2 K (total wall only 0.131 W/m^2 K) (Fig. 6.14).

The following illustration shows the construction outlined above including a window frame. The window frame would have to be adapted to suit the design in Sect. 5.3.

6.8.2 Interior Walls

All interior walls on the ground and upper floors are insulated as standard and panelled with OSB board on both sides. The total thickness of the interior walls is approximately 12.5 cm.

Interior walls are can be made as structurally strong as exterior ones. Shelves and cupboards weighing up to 250 kg can be attached anywhere on the wall, using ordinary screws. The bonded construction holds the thermal and sound insulation material permanently in place, preventing slippage. Cable channels are integrated inside the wooden frame, meaning that the wiring can be extended later (Fig. 6.15).

Fig. 6.15 Hanse Haus detail for interior walls

Partition walls, interior walls on the upper storeys and walls that deviate from the standard clear room height (approximately 2.5 m), are constructed slightly differently from the exterior, gable and interior walls described below.

6.8.3 Ground Floor Construction

Moisture barrier (with floor slab).
130-mm thermal insulation (PUR).
Underfloor heating on heat-reflective aluminium foil.
55-mm wet screed for underfloor heating.
The thermal insulation value of the floor on the lower storey (in houses with floor slabs) is 0.16 W/m^2 K.

6.8.4 Upper Floor

12.5 mm	fire-protective plasterboard.
60/30 mm	substructure (battens).
240 mm	load-bearing solid joists.
100 mm	thermal and sound insulation (mineral wool insulation).
18 mm	OSB board, complete coverage.
60 mm	thermal insulation (polystyrene).
40 mm	impact sound insulation (polystyrene).

Underfloor heating on heat-reflective aluminium foil.

55 mm Wet screed for underfloor heating.

6.8.5 Gable Roof

Concrete roof tiles in a range of colours
Battens
Counter-battening
Vapour-permeable sarking membrane for the highest possible protection against wind-driven snow and dust
240-mm rafters, depending on structural requirements
240-mm thermal and sound insulation (mineral wool insulation, WLG 035) between the rafters
Vapour barrier
60/30-mm substructure (battens)
12.5-mm fire-protective plasterboard

The thermal insulation value of a Hanse Haus gable roofs is 0.16 W/m^2 K.

Roof cladding in high-quality concrete roof tiles comes with a 30-year manufacturer's performance guarantee.

6.8.6 Windows and Doors

Thermally efficient six-chamber plastic profiles reinforced with integrated steel, creating a highly stable frame.

Triple glazed to reduce heat loss, giving a thermal insulation value of just 0.5 W/m^2 K.

The edges of the individual panes are thermally isolated from each other, preventing the build-up of condensation.

Lockable handles on all windows and patio doors.

6.8.7 Heating—Underfloor

Vaillant gas condensing boiler, which runs on natural gas, uses the latest condensing boiler technology. The integrated high-efficiency pump means there are savings of up to 15% on heating and up to 50% on electricity bills when compared with traditional non-condensing boilers.

A 200 L Vaillant hot water cylinder with an air-to-water heat pump externally provides both heating and hot water. The system is exceptionally cost-efficient, because its modulating compressor continually adjusts to the current heating load in the home. The gas condensing boiler comes with an outdoor temperature control system: a central adjustment module uses intelligent technology to make individual, user-determined adjustments and carry out checks. The heating system will be installed in the utility room.

6.8.8 Ventilation

Ventilation in the living areas keeps out dust and pollen while routing CO^2 and damp outside. An efficient, integrated heat recovery system (MVHR) transfers the heat in the waste air to the fresh air streaming in, without the two mixing. Via a system of pipes and air vents, the heated fresh air is distributed to the individual rooms. It includes a filtration system as well as supply and exhaust valves and a system control panel.

6.8.9 *Electricity*

Supplied by small wind turbine or electricity from a 'green' supplier. Photovoltaic panels are often used in combination with Air Source Heat Pump (ASHPs).

6.9 Conclusion

I do not claim to have presented a total coverage of the technology of house building but instead I have selected those aspects that seem relevant to my focus of study. This book is not a 'how to' book with regard to building a house but rather it is intended to be a stimulant towards more consideration for the design aspects with regard to the interior facilities provided. The main enemy to any sort of progress is tradition when faced with change, misguided institutional profit motives, and poor advice to government and local authorities. It is time for a shakeup in the industry and a complete review of space standards is desperately required. It does not help when the teaching in our schools of architecture and design, judging by recent graduates' work, seems to steer students into the grand visions associated with modernistic principles, which is socially divisive. Instead of revolution, we have rather meek acquiescence with the status quo and the production of an architecture devoid of any humanity. But then we are beginning to see humanoid robots being made so maybe they will fit in with the new architecture perfectly.

Acknowledgements: www.vaillant.co.uk
www.worcester-bosch.co.uk
Hanse Haus GmbH & Co.
Building Research Establishment (BRE)
The National Self-Build & Renovation Centre
www.greenmatch.co.uk
Hafele
Halliday and Baillie
ECOHAUS Company
UK ROLLER SHUTTERS
Mantle Building System
StrideTreglown—Rob Delius

Bibliography

Baker, Lisa. 2016. *The Sub/Urban Idea: From Terraced Houses to Residential Estates*. Braun.

Chung-Klatte, Sharon, Ulrich Knaack, and Reinhard Hasselbach. 2010. *Prefabricated Systems*. Basel, Switzerland: Birkhauser Verlag AG.

Clarke, Alison J. 2017. Editor of *Design Anthropology*. London and New York: Bloomsbury.

Corbusier, Le. 1946. *Vers Une Architecture (Towards a New Architecture)*. Jean-Louis Cohen, Architectural Press.

Crosby, Theo. 1973. *How to Play the Environment Game*. Harmondsworth: Penguin Books Ltd.

Davies, Colin. 2005. *The Prefabricated Home*. London: Reaktion Books Ltd.

Department of Education (DOE). 1982. *Activities and Spaces*. Architectural Press. December.

Dorling, Danny. 2014. *All That is Solid*. UK: Penguin Books.

Downey, Georgina. 2013. *Domestic Interiors*. London: Bloomsbury.

Greenwich Council. 2007. *Progress Estate Conservation Area*.

Historic England, (Formerly English Heritage). 2011. *Domestic 4, The Modern House and Housing*. Listing Selection Guide. April.

House for the Future, National Museums and Galleries of Wales. Museum of Welsh Life St Fagans, Cardiff.

Hunt, Jamer. 2018. *Design Anthroplogy*. Editor Alison J. Clarke. Bloomsbury Academic.

Knaack, Ulrich, Sharon Chung-Klatte, and Reinhard Hasselbach. 2010. *Prefabricated Systems*. Birkhauser Verlag AG.

Leupen, Bernard (ed.). 2011. Architect Harald Mooij, *Housing Design—A Manual*. NAI Publishers.

Mayor of London. 2010. *London Housing Design Guide*. London Development Agency.

McGrath, Paul. 2015. *Why Space Standards Are a Bad Thing For the Housing Crisis*. BD Online. January 5.

Mental Health Foundation, Colechurch House, 1 London Bridge Walk, London, SE1 2SX.

Meuser, Philipp. 2017. *Prefabricated Housing: Construction and Design Manual*. DOM Publishers.

Minton, Anna. 2017. *Big Capital*. UK: Penguin Books.

Nash, Clare. 2016. *Contemporary Vernacular Design*. London: RIBA Publishing.

O'Brien, Charles. 2016. *Houses: An Architectural Guide*. London: Yale University Press.

Parissien, Steven. 2007. *Interiors, The Home since 1700*. London: Laurence King Publishing.

Park, Julia. 2017. *One Hundred Years of Housing Space Standards*. January. housingspacestandards.co.uk.

Per, Aurora Fernández, Javier Mozas, and Javier Arpa. 2011. *A&t Research Group, Density is Home*. Spain: a+t Architecture Publishers.

Rapoport, Amos. 1969. *House, Form and Culture*. New Jersey: Prentice-Hall.

Ross, Keith, of the BRE. 2002. *Non-Traditional Housing in the UK*. July.

Schneiderman, Deborah. 2012. *Inside Prefab*. Architectural Press.

A. Sully, *The Estate House Re-designed*,
https://doi.org/10.1007/978-3-319-90397-2

Sowerby, Githa. 1907. Illustrated by Millicent Sowerby. '*Childhood*' Published by Chatto and Windus.
Steadman, Philip. 2006. Why Are Most Buildings Rectangular? *Architectural Research Quarterly* 10, no. 2.
Stevenson, Greg. 2003. *Palaces For the People/Prefabs in Post-war Britain*. London: Batsford.
Wentling, James. 2017. *Designing a Place Called Home*. Switzerland: Springer.
The Housewife's Book published by the Daily Express.
The Modern House. London, N1 3HF.
www.whathouse.com/awards/.
www.socialhousinghistory.uk.
www.insidehousing.co.uk.

Reports/Papers

Design Research Society. 50th Annual Conference Proceedings June 2016. Future Focused Thinking.
Estate Regeneration Forum. NHBC. The Housing Forum. January 2017.
Historic England. Domestic 4—The Modern House and Housing. April 2011.
House for the Future. St Fagans. National Museums and Galleries of Wales. 2001.
House of Commons Hansard. Quality in the Built Environment. December 13, 2017.
London Housing Design Guide. Mayor of London. London Development Agency. August 2010.
More homes, fewer complaints. Commission of Enquiry into the Quality and Workmanship of New Housing in England.
Non-Traditional Housing in the UK. Building Research Establishment. 2012.
Parker Morris Report 1961.
RIBA. *Home Improvements: Housing Research in Practice*. October 2013.
RIBA. *Improving Housing Quality*. 2009.
RIBA. *Case for Space Report*. June 2017.
RIBA. *Space Standards for Homes*. Homewise. December 2015.
RIBA. *RIBA Home Improvements: Housing Research and Practice*. October 2013.
RIBA. Hodder, Stephen. RIBA President.
RIBA. *Starter Homes Design*. Department for Communities and Local Government. March 2015.
RIBA. *What House?* Awards 2016.

Organisations

Design for Homes. www.designforhomes.org.
Home Builders Federation, HBF House, 27 Broadwall, London, SE1 9PL.
Lifetime Homes. www.lifetimehomes.org.uk.
Mental Health Foundation. www.mentalhealth.org.uk.
Mental Health Network. www.nhsconfed.org.
National Housing Federation. www.housing.org.uk.
NHBC (National Housebuilding Council), NHBC House, Davy Avenue, Knowlhill, Milton Keynes, Bucks, MK5 8FP.
Royal Institute of British Architects.
The Built Environment Trust. www.buildingcentre.co.uk.
The Chartered Institute of Building. www.ciob.org.
The Chartered Institute of Housing. www.cih.org.
The Design Commission. www.policyconnect.org.uk.

Index

A. Sully, *The Estate House Re-designed*,
https://doi.org/10.1007/978-3-319-90397-2

MIX
Papier aus verantwortungsvollen Quellen
Paper from responsible sources
FSC® C105338

If you have any concerns about our products,
you can contact us on
ProductSafety@springernature.com

In case Publisher is established outside the EU,
the EU authorized representative is:
Springer Nature Customer Service Center GmbH
Europaplatz 3, 69115 Heidelberg, Germany

Printed by Libri Plureos GmbH
in Hamburg, Germany